Slope Earthquake Stability

Yang Changwei · Zhang Jingyu
Lian Jing · Yu Wenying · Zhang Jianjing

Slope Earthquake Stability

Science Press
Beijing

Springer

Yang Changwei
Southwest Jiaotong University
Sichuan
China

Zhang Jingyu
Southwest Jiaotong University
Sichuan
China

and

Oklahoma State University
Stillwater, OK
USA

Lian Jing
Southwest Jiaotong University
Sichuan
China

Yu Wenying
Southwest Jiaotong University
Sichuan
China

and

Oklahoma State University
Stillwater, OK
USA

Zhang Jianjing
Southwest Jiaotong University
Sichuan
China

ISBN 978-981-10-2379-8 ISBN 978-981-10-2380-4 (eBook)
DOI 10.1007/978-981-10-2380-4

Jointly published with Science Press
ISBN: 978-7-03-049541-9 Science Press, Beijing

Library of Congress Control Number: 2016947933

Printed on acid-free paper

This Springer imprint is published by Springer Nature
The registered company is Springer Science+Business Media Singapore Pte Ltd.

Preface

Researching on the dynamic stability of slope subjected to earthquake is an important project in geotechnical engineering field. However, the engineering geology, soil mechanical, geophysics, and earthquake engineering is involved in this project, which is very difficult, and now there are many problems which must be overcome. Both the special geographical conditions and geologic structures determine that the problem of seismic stability of slope is very conspicuous in China. However, the infrastructures in China is basking in a great boom such as road, railway and water conservancy, which needs higher demands for the seismic design of slopes and the hazard assessments for landslides , and the existing theories and practices can't satisfy this need. Therefore, researching on the seismic stability of slopes is very important, which is useful to reduce the hazards caused by the landslides as soon as possible, produces a favorable living environment for the people living in the mountain area, and ensure the normal and safety operation of the geological engineering facilities.

The content of this book can be divided into four parts, which are written by five authors that Yang Changwei, Zhang Jingyu, Lianjing, Yu Wenying, Zhang Jianjing. The concrete content are introduced synoptically, as follows: Based on the simply review of seismic stability of slopes and the background of Wenchuan earthquake, aiming at the related science problems existing in the current phase, following the dynamic characteristics of slopes subjected to the earthquake, using the methods such as the field investigation, numerical analysis, shaking table test, the monitoring data of the seismic array and the theoretical derivation, taking the covering layer- bedrock type slope which is destroyed seriously, this book reveals that dynamic characteristics of the rock slope, and analysis the formation mechanisms of the landslides, and then proposes the assessing system of the landslide hazards and a method which can consider the three factors of earthquake wave which are duration, amplitude and frequency, respectively.

In the process of compiling this book, thanks to Researcher Cuipeng, Professor Huang Runqiu, Professor Yao Lingkan for their help with this article. Thanks to Doctor Fu Xiao, Liu Feicheng and so on for assisting in collating materials. At the same time, this book is supported by National Science Foundation (No. 51408510) and Science and Technology Project of Ministry of Communications NO. 20133 18800020).

This book cites some research results of our research team (such as the Chinese book named THREE DIMENSIONAL SPACE-TIME ANALYSIS THEORY GEOTECHNICAL EARTHQUAKE ENGINEERING). A lot of other researchers' results are referred and cited. Authors list the references in the book as comprehensive and complete as possible, but it is hard to omit some references. If there are some mistakes about the references or the other content, please forgive and correct us. Thank you very much!

Sichuan, China — Yang Changwei
Sichuan, China and Stillwater, USA — Zhang Jingyu
Sichuan, China — Lian Jing
Sichuan, China and Stillwater, USA — Yu Wenying
Sichuan, China — Zhang Jianjing
November 2016

Contents

Chapter 1
Introduction

In 2008, Wenchuan earthquake (Ms8.0) was triggered by Longmen Shan fault, causing more than 87,000 people killed and 370,000 wounded. Wenchuan earthquake is the most destructive earthquake occurred in China inland in the past 100 years. Two years after, Yushu, Qinghai Province, was struck by earthquake with a magnitude of 7.1, inducing over 2000 deaths. Natural disasters pose lots of scientific issues to be addressed.

Seismic stability of slope is one of these urgent issues induced by earthquake disaster. Revealing the mechanism of earthquake-induced landslides, assessing the seismic stability of slope, and forecasting the prospective hazard region are the common goal of geotechnical engineers, earthquake engineers, and seismologists.

1.1 Background and Significance

Earthquake geological disaster is one type of geological phenomenon triggered directly or indirectly by the force of inertia; this disaster can cause damage to human life, property as well as environment, including landslides, collapse, mudslides, and liquefaction of saturate sand.

Earthquake damage survey shows that earthquake-induced landslides are the most common geological disaster in earthquake-inflicted areas with large number, wide distribution. Keefer studied 40 slumps globally; his result shows that slumps will be triggered by earthquakes with a magnitude no less than 4.0. The number of geological disasters induced by Wenchuan earthquake exceeds 20 thousand, the majority of which is landslide and collapse. The number of massive landslides with sliding volume more than $1000 \times 10^4\ m^3$ is over 30, as shown in Fig. 1.1. Daguangbao landslide in an County is the largest earthquake-triggered landslide worldwide, causing more than 0.7 billion sliding body. Chichi earthquake (1999), Taiwan, also triggered massive landslides and collapses. Izmit earthquake (1999),

Y. Changwei et al., *Slope Earthquake Stability*,
DOI 10.1007/978-981-10-2380-4_1

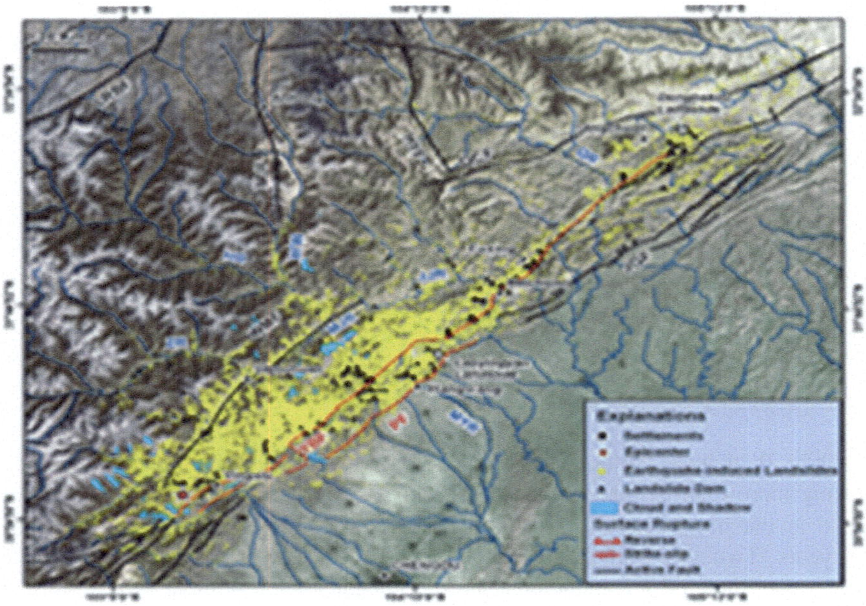

Fig. 1.1 Distribution map of collapse and landslide hazard points in Wenchuan earthquake

Turkey, caused large-scale surface rupture and destroyed lots of highway embankment cutting slopes in mountain areas.

The earthquake-induced landslide caused serious consequences, like clogging river, block communication, and other lifeline projects. Diexi earthquake (Sichuan Province, China, 1933) triggers rock collapses which block the Minjiang River, causing water level rises 300 m at Yinpin Cliff and dam collapse in two month later, flood swept both sides of the Minjiang River, inducing a total of 20,000 deaths. Traffic disruption caused by earthquake will greatly slows the rescue effort. Wenchuan earthquake (Sichuan Province, China, 2008) blocked 7 highways, 5 national roads, and 10 provincial roads, which hindered rescue work in earthquake-strike zone, as illustrated in Fig. 1.2. Dujiangyan to Yingxiu section of national road G213 (30.6 km long) was inflicted by 45 earthquake-induced landslides and 2 debris flows, and 1770-m-long road was buried.

China is a mountainous country, mountainous area accounts for about two-thirds of Chinese land area, which formats a large number of man-made and natural slopes. China locates between circum-Pacific seismic belt and the Eurasian seismic zone, lots of earthquake fault zone are well developed. Hence, China is an earthquake-inflicted country. Seismic stability of slope is critical to the construction of buildings and infrastructures in high-seismicity area.

Slope stability analysis under dynamic loads is still in the exploratory stage. Systematically investigations to the dynamic stability of slope, its failure mechanism, stability assessment method, and hazard range prediction will be meaningful both theoretically and practically.

Fig. 1.2 Landslide in miles of K1034+200 ~ 280 on the left side of G213

1.2 Review

Slope stability is significant in geotechnical engineering research. Lots of studies have been undertaken by researchers throughout the world in the past century, and useful results have been obtained. The evolution the slope has been recognized gradually and the evaluation of slope stability analysis is also developed from the initial qualitative analysis to quantitative. In general, slope static stability analysis has matured, and corresponding technical specifications have been made to guild engineering practice. However, since the complexity of research on the slope dynamic stability analysis, the seismic response, and stability analysis of slope should be explored further in much more details systematically.

1.2.1 Shaking Table Test Technology

Shaking table simulation technology is the most useful and direct method to study the seismic performance of structure, seismic response of geotechnical material, and the mechanism of geological disaster. The rules underlying the earthquake disasters can be revealed through shaking table test; hence, shaking table test is one of the most significant ways to undertake earthquake engineering study.

Shaking analog technology began in the 1960s, Japan is the first country in the world to research and establish shaking table technology. Japan currently has the world's largest shaking table (15 m × 15 m) which can vibrate horizontally and vertically at the same time. The first unidirectional oscillation shaking table (3.65 m × 3.65 m) was established in America in 1968, later a bidirectional shaking table (6.1 m × 6.1 m) was established in UC, Berkeley.

In China, research on shaking table technology began in the 1960s, developed from the 1980s to the 1990s. The first six-degrees-of-freedom shaking table simulation station was established in China in 1997. A team led by Prof. Bolong Zhu set up a bidirectional shaking table in Tongji University, and then, multiple transformations have been done to this shaking table system, upgrading its carrying capacity from 15 to 25 t, as well as this control system and data acquisition system. Distinctive shaking tables have been established in China Academy of Building Research, Harbin Institute of Technology, Beijing University of Technology, Institute of Engineering Mechanics, etc. (Figs. 1.3 and 1.4).

So far, shaking table simulation has been given broad application to construction, harbor, water conservancy and hydropower, railways, agriculture, and other sectors. Shaker not only be able to detect seismic performance and but also can carry out geotechnical engineering, mechanical, and electrical products, inspection, and test human reactions. According to its load capacity and table size, vibration table can generally be divided into three categories: small-size shaking table with table size less than 2 m × 2 m and load capacity no more than 10 t; medium-size shaking table with carrying capacity of 20 t or so and table size between 2 m × 2 m and 6 m × 6 m; large-size shaking table with load capacity of more than 20 t. Currently,

Fig. 1.3 Multi-axial electrodeless adjustable array system

Fig. 1.4 Shaking table with the size of 6.1 m × 6.1 m

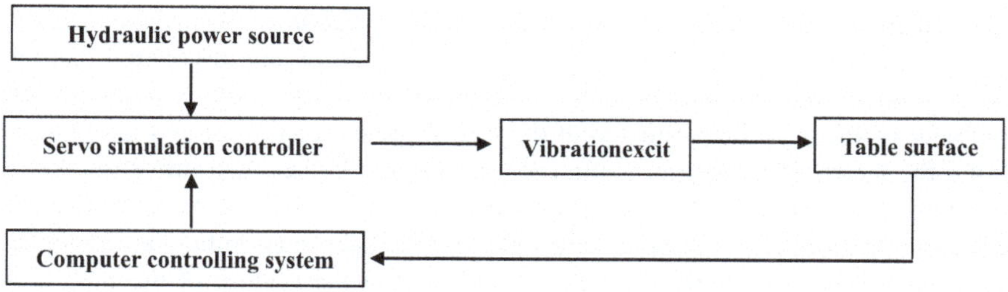

Fig. 1.5 Shaking table system of seismic simulation

the most common shaking table is electric–hydraulic servo-vibration table which comprises of hydraulic pressure source system, shaker, analog servo-controllers, countertop, and computer control system. Servo-hydraulic valve analog controller is the core of the analog controller and determines the performance the whole shaking table system. Hydraulic system is mainly to provide power source, including pump stations, accumulators, cooling systems. Pump flow is designed according to the maximum value of seismic wave velocity. In order to save energy, large capacity accumulator group is set to provide the actuator tremendous energy instantaneously, as shown in Fig. 1.5.

1.2.2 Dynamic Characteristics of Slope

The dynamic characteristics of a slope under earthquake refer to the seismic response of acceleration, velocity, displacement, stress and strain, etc. It relates not only to the incident motion but also to the internal characteristics of geotechnical body, complicating the dynamic response compared with statistic analysis. Earthquake disaster investigation, theory exploration, numerical method, and model test are employed to conduct this research.

1.2.2.1 Filed Investigation

Earthquake left not only disasters, damage, and risks but also many valuable repository. This valuable repository can provide a reference for subsequent disaster prevention work.

A database of seismic landslide hazard has been established by Keefer in 1984, the relationship among landslide distribution, lithology, slope angle, magnitude, and epicenter distance was analyzed by Keefer based on the survey data of earthquake Loma Prieta. It concluded that earthquake-induced landslide density increases with increasing slope gradient but decreases rapidly with the increase of epicenter distance. Rodriguez statically analyzed the seismic landslide hazard data during 1980–1997 all over the world and concluded that most of the landslide occurred in intensity zone of 7 and 8 degrees. Sepulveda et al. investigated the Pacoima Canyon area after Northridge earthquake in 1994 and discovered that the acceleration of the bottom the canyon is less than 0.5 g while that at the left abutment of Pacoima dam situated at the top of the canyon is up to 1.6 g; meanwhile, the landslide density near the canyon is much higher than the surrounding area, which indicated that the topography has an crucial effect on the amplification of seismic motion. Davis et al. discovered that the shaking amplitude at the top of mountain has been multiplied compared with that at the bottom of the mountain after their observation to the aftershock of San Fernando earthquake. Hideo Takano et al. found that seismic intensity on the slope increased by about 1 degree relative to the bottom of the slope. After the research on measured seismic data, Celebi discovered that damage caused by earthquake tends to intensify on the ridge and cliff, which indicating the topographic simplification effect of seismic waves.

In China, based on earthquake-induced natural slope slump case in the past nearly 800 years, Hongxin Bo et al. proposed prediction method for the earthquake-induced slump, pointed out the minimum magnitude-triggered landslides and the earthquake-induced slope collapse mainly occur within the epicenter distance of 100 km. Shaowu Sun has researched the distribution of collapse and landslide triggered by historical earthquakes and concluded that collapse landslides caused by the earthquake mainly concentrated in strong earthquake-prone and terrain complex zone. Yanhui Yu et al. proposed a grading prediction method applicable to Western China. Shuhua Cai et al. developed a method to calculate the largest sliding distance of earthquake-triggered landslide, established a quantitative relationship between the magnitude of the earthquakes and the largest landslides slipping distance.

The Wenchuan earthquake-induced landslide characteristics were analyzed by Yin and Peng et al. and concluded that landslide sliding bed often do not have a continuous full slip surface. Xie and Qiao studied the distribution of earthquake-induced landslide; discovered that secondary mountain hazards majorly distributed along Longmen Shan fault zone; analyzed the relationship among the landslide distribution, seismic intensity, and macroscopic epic central distance; obtained landslide hazard area and density for the study area. Huang established a classification system to the mechanism of earthquake-triggered landslide, the effect of slope failure mechanism under strong earthquake is divided into five categories; slip collapse, collapse, projection, peeling, and shatter, revealing the unique dynamic characteristics of Wenchuan earthquake-triggered landslides.

After the Wenchuan earthquake, the author rushed to the earthquake disaster area and investigated the distribution and scale of landslide as well as the seismic response of supporting system.

1.2.2.2 Model Test

Model test has been proved to be an effective means to simulate the seismic response of geotechnical body. Model test can reflect the progressive failure mechanism and stability of the slope of the soil truly and visually and provides an important basis for a variety of numerical simulation results.

Meei-Ling Lin et al. conducted large-scale shaking table test to homogeneous sand, the seismic response of model to motions with different frequencies and amplitudes was obtained. The slope presented a linear response under amplitude less than 0.4 g, and the nonlinear response began when amplitude exceeded 0.45 g. Yung-shah Hong et al. tested a total of 5 test to nail-supported slopes. The effects of soil nail inclination, length, and frequency of the input seismic wave on the seismic performance of slope were investigated. Results show that soil nailing significantly improve the seismic performance of the slope, nail inclination hardly affect the slope deformation and prolonging soil nail length can significantly improve the seismic performance of the slope.

Wang, Juming Zhang, and Xiangong Zhang et al. undertake small-scale shaking table test to rock to explore the friction characteristics of slope surface and established a three-dimensional block movement of kinetic differential equation. Feng et al. [52] tested the seismic response of both single-sided and double-sided slopes and considered the role of earthquake seismic waves, causing the major changes of stress field in slope body; in particular, the magnitude and direction of tension and shear of the slope appear to repeatedly change. Qiang Xu et al. discovered that inertial force triggered by earthquake is the main cause of slope deformation and failure.

The author's research group conducted shaking table test to gravity retaining wall, computational model to the displacement of gravity retaining wall. Large-scale shaking table test was also conducted to study the effect of ground conditions on the seismic active earth pressure of retaining wall. Large-scale shaking table test to anchored pile wall has also been done to explore its seismic performance and seismic design optimization method.

1.2.2.3 Numerical Simulation Method

Numerical simulation can be used to study the seismic characteristics of slope from both microscopic and macroscopic perspective; numerical method also can address the issues encountered in shaking table test, such as non-conformance similar system and size effect, and expand the results of the shaking table test.

Since finite element method has been introduced in geotechnical dynamic analysis by Clough and Chopra and the advance in computer technology, finite element method played a huge role in slope seismic analysis. Finite element method can be applied not only to the total stress method but also to effective stress method. Finite element method can consider complex effects of topography, nonlinearity and heterogeneity, and be able to analyze the self-vibration characteristics and dynamic response of the various parts of a geotechnical body, which enables finite element method becomes one of the most powerful methods for dynamic analysis of geotechnical body.

Jianliang Huang et al. employed Sarma method to carry out dynamic analysis of seismic stability theory, derived the critical acceleration formulas for slope in both horizontal and vertical directions, and established slope stability estimation process by using slice technique for any form of slope in accordance with the horizontal and vertical seismic acceleration curve.

Shengwen Qi et al. used Flac 3D to obtain the general rule of the dynamic response of slope displacement, velocity, and acceleration; two different slope dynamic response patterns were found. Laigui Wang based on finite element method to simulate the earthquake-induced slump of single weak surface slope. Shunyi Wu et al. used the particle flow code (PFC) to simulate the failure process of soil slope with weak interlayer.

Changwei Yang et al. utilized a novel numerical code continuum-based discrete element method (CDEM) to analyze the dynamic characteristics of mountain valleys under earthquake and concluded that mountain terrain has a significant impact on the amplification to both horizontal and vertical acceleration, while to the valley terrain, the acceleration amplification effect has been restrained in a certain range.

1.2.2.4 Theoretical Analysis

Research on rock slope mainly remains on field observations, shaking table test, and numerical simulation. As few studies on theoretical solution of seismic properties of rock slope have been done and cannot be used to guide the engineering practice, further in-depth study is urgently needed.

Based on the elastic wave theory, Kurita, R.R., and Rodriquez-Ovejero derived the theoretical solution of sided rock slope to harmonic waves.

Chong Shi used plane ray theory derived from the theoretical solution for single slope elevation amplification effect under earthquake, which can be utilized to describe the reflection of invading slope face to seismic waves. Elevation amplification effect of slope is due to the reflection of the free surface of a slope. The coupling effect of dynamic elastic modulus, Poisson's ratio, the slope, and the incident frequency are present in a sparse elevation amplification factor. The higher the incident frequency, the more obvious of the rhythm is and the more extreme points. The slope may bear a much more intense seismic response compared with the toe because of the free surface reflection, interface reflection, and refraction.

1.2.3 Deformation Characteristics and Instability Mechanism

Earthquake-induced landslides are important geological and seismic hazards; its complex and diverse mechanism makes the prediction much more difficult. Hence, slope slide is one of the most difficult geological problems around the world.

Earthquake slope instability mechanism is the key problem that must be settled. Researches on its deformation characteristics and slope instability mechanism under the earthquake are essential. So far, abundant research results have been obtained; representative results are as follows:

W.H. Zhou indicated that the seismic stability of rock slope is determined by the reflection or refraction when seismic waves travel across the interface between layers or the surface as well as the interaction of earthquake loading with other factors (pore water pressure); Hartzell studied the Robinwood Ridge destroyed by Loma Prieta earthquake and then discovered that the main factor causing the destruction of the ridge includes the reflection and transmission of waves with the ridge, complex role of the Rayleigh wave and Love wave, main source directivity and wave diffusion state. G.T. Cui proposed slope fluctuations oscillation acceleration effect hypothesis. R.L. Mao believed that the fluctuation oscillation of slope produces three effects: progressive destructive effect, triggering effect, and accelerating effect. S.W. Qi et al. think that the instability of slope is caused by both the effects of seismic inertial force and the build-up of excess pore water pressure. C.A. Tang et al. consider that incident principle of stress wave in free surface and acceleration multiplier effect can explain the mechanism of strong ground motion and numerically simulated the crack and projectile phenomenon in slope surface. F.P. Cui et al. considered that the formation of slump is caused by the time difference coupling effect of seismic P waves and S waves. Q. Xu et al. tested the seismic response of slope by shaking table test and numerical method and then proposed that the internal and essential failure-mode large-scale earthquake-induced landslide can be summarized by crack slip, dominantly developed crack slip along strike slope, crack horizontally slip, crack dispersion slip.

In summary, regardless of the existing results on the earthquake-triggered landslide, further study is urgently needed.

1.2.4 Seismic Stability Evaluation Method

Static slope stability issues-related research has matured so far; lots of states have developed corresponding technical specifications for engineering practice. However, because of its high degree of complexity, researches on the dynamic response and stability of the slope under earthquake are far from mature.

Domestic and foreign scholars have evolved a variety of seismic response analysis method of slope, assessment methods of slope stability can be divided into two categories: deterministic and probabilistic analysis methods from an uncertainty perspective. From the different treatments of slope stability calculations, assessment methods of slope stability can be divided into four categories: ① pseudo-static method, ② slice method, ③ numerical method, and ④ experimental method. Hereinafter, the pseudo-static method, slice method, numerical method, and experimental method will be discussed in detail.

1.2.4.1 Certainty Analysis Method

① Pseudo-static method

Since its first application to slope seismic stability analysis, this simple and practical method had been widely employed. This method has been included in railways, roads, and water and other related design specifications. Based on this method, a series of research results on the seismic slope stability analysis have also been achieved by experts and scholars. The most critical issue is to select the seismic coefficient. Seed summarized the three determining seismic coefficient selection methods: experimental method, rigid body reverse analysis method, and viscoelastic response analysis method; the experimental method is the most widespread used one among these methods.

Seed used force polygon method to calculate the static stress on the slip arc with consideration to earthquake inertial forces. Stability safety factor was calculated. Leshchinsky et al. evaluated slope stability by pseudo-static method, applied numerical method to calculate the normal stress on the sliding surface, determined the minimum safety factor for all limit equilibrium equations, and finally proposed a table for slope seismic stability assessment. Bray et al. used wave propagation theory and pseudo-static method to analyze landfill site with weak layer. F.B. Zeng developed a slope stability calculation method which considered the soil internal friction angle affected by strength of the earthquake on the basis of friction circle method proposed by Taylor. Ling et al. applied the pseudo-static method to analyze the seismic stability of rock body sliding along the surface of the rock joints, and calculated the permanent displacement. Siyahi et al. carried out seismic stability analysis of slope with normally consolidated soil by using pseudo-statistic method and determined the safety factor under different shear strengths. Pseudo-static method was applied to the seismic stability analysis of reinforced slope by Ausilio et al. and considered different failure modes by the use of limit theory, proposed calculation formulation for reinforcement force. Biondi et al. studied the effect of pore water pressure on the stability of infinite saturated clay slopes during earthquake and after the earthquake based on pseudo-statistic method. Siad combined unlimited design theory and pseudo-static method to derive the coefficient formula. P.F. Lv improved the pseudo-static method, the horizontal slice was considered while calculating the horizontal seismic force, and the vertical slice, while vertical seismic force.

While the pseudo-static method is simple, its obvious deficiencies exist. During actual earthquake, the value and direction of the seismic inertial forces are rapid fluctuations over time. During an earthquake, even though the slope stability factor is less than 1 temporarily, unstable failure will not necessarily happen, only produce a certain permanent displacement; the stability of slope is controlled by the magnitude of the deformation instead of stability factor derived form pseudo-static method. Z.J. Shen indicated that fundamental flaw of pseudo-static method is unable to consider the spectral characteristics of ground motion and impact of duration.

② Slice Analysis method

Pseudo-static method only provides a stability index (Safety factor), rather than information on the failure surface deformation. Limited slip displacement of thought was first proposed by Newmark in the Fifth Rankine Lecture, 1965. He pointed out that the slope stability depends on the deformation caused by earthquakes, not minimum safety factor. Due to the short time of the earthquake, the inertial force changes direction back and forth. Even if the inertial force is sufficiently large during a short period of time, which makes potential landslide safety factor less than 1.0 and leads to permanent deformation of slope, as soon as acceleration is reduced or even reversed, the slide stops, slope will not produce further permanent displacement if the soil does not significantly reduce its strength after earthquake. The basic thought of the Newmark method showed that permanent displacement slope is the most important parameter for slope stability analysis under earthquake.

Since the Newmark limited slip displacement method has been proposed, the domestic and foreign scholars have been highly concerned about this method and conducted in-depth researches, and the Newmark method has been widely applied in practice. Crespellani T. introduced earthquake damage trend factor (PD) to consider the main factors that control slope stability. By analyzing the 310 real horizontal seismic waves, Crespellani T. proposed empirical formula between the Newmark rigid slider displacement and PD. Based on the Newmark method, Liangzhi You et al. calculated the permanent landslide displacement of homogeneous soil slope by using different ground motion records and then presented the chart that showed the relationship between the different yield acceleration and permanent displacement. T. Crespellani et al. investigate the effects of filtering amendment to Newmark permanent displacement. S.J. Wang introduced the thought of limited slip displacement method to dynamic analysis of slope stability and proposed dynamic method for slope block sliding. S.J. Wang developed the concepts of movement initial friction and movement friction through tests and measured the relation ship between dynamic friction coefficient and velocity on smooth joints surface of granite. Based on the results of S.J. Wang, WANG Si-jing, Xueshou Yi, and Zhang Juming derived the three-dimensional dynamic reaction equations of wedge-shaped and layered mountain. A simple method is proposed to calculate the seismic force of permanent displacement by S.L. Qi et al. and this method was based on the residual thrust method and the Newmark method and considered the changes of pore water pressure caused by dynamic force.

Despite the widespread application of the limited slip displacement method in engineering practice, this method fails to consider the cumulative deformation of slip zone and only can provide approximate sliding deformation with poor accuracy.

③ Numerical Simulation method

So far, finite element method, discrete element method, and the fast Lagrangian method have been widely employed in slope stability analysis of earthquake slope

stability assessment index includes safety factor and permanent displacement, as discussed hereinafter:

In terms of the safety factor combined with limit equilibrium theory, some valuable results have been obtained. J.S. Bo et al. generalized the dynamic stress state of soil slope as the superposition of gravity stress and additional dynamic stress, combining time-domain concentrated mass fluctuations explicit finite element method and multiple transmitting formula to analyze the fluctuation of dynamic displacement field, pressure field, and the stability factor. L.L. Chen et al. established a formula for calculating the stability of high steep rocky slope; dynamic response was acquired by response spectrum and time history analysis method, and the shear strength reserve of potential slide surface was presented. The results can be used to evaluate slope stability. Z.Y. Wu et al. calculated the time history of safety factor by finite element method; time-weighted average safety factor was given. Chen Jian et al. studied the stability of a permanent ship lock steep slope in the Three Gorges by using pseudo-static method. It is noteworthy that the above methods are based on the assumption of small deformation, failed to take into account the impact of the large deformation, vertical ground motions are rarely studied.

As for permanent displacement, valuable research results on finite element method have been achieved. Permanent displacement can be obtained by integrating permanent strain of each element in finite element mesh. Calculation methods can be categorized to strain trend method, stiffness reduction method, and nonlinear methods. Seed et al. proposed strain trend method according to the linear and equivalent linear analysis results in 1971. Lee and Serf proposed stiffness reduction method. C.H. Zhang et al. studied the influence of ground motion input on the permanent displacement of the Three Gorges ship lock and high rock slope, and the results were compared with those from the pseudo-static method. L.J. Tao et al. calculated the permanent displacement of a rock slope by discrete element method and evaluated its stability. Y.M. Chen et al. proposed a permanent displacement calculation method based on circular slip and Newmark method.

Both slope safety factor and permanent displacement can be calculated by finite element method. Numerical simulation is often related to computer technology, complex programming, and it is time-consuming. In particular, for in-homogeneous rock slope, it is difficult to be promoted in engineering practice.

④ Experimental method

With the continuous development of shaking table technology, the number of vibration table is increasing rapidly. Researches on the seismic stability of slopes based on vibration table test are gradually increasing. Shaking table test has become one of the main means of seismic rock slope stability study. A series of promising research results have been made by C.W. Yang et al., Q. Xu et al., Y.M. Men et al., H. Ye et al., and W.K. Feng.

Although seismic stability test method can present intuitive response of slope, the shaking table test is costly, time-consuming, and the experiments similar rate as well the choice of materials issues have not been addressed yet.

1.2.4.2 Probability Analysis

There are many uncertainties in seismic slope stability analysis, like the randomness of input ground motion and material properties. Only reasonable consideration to the randomness of these parameters is paid, disasters level can be clearly described. Therefore, it is essential to develop the probability method. As the study in this area was carried out late, only limited reviews are available.

Lin et al. studied the probability of failure of the slope based on the hypothesis which regarded the strong ground motion as a Gaussian process. Hisakzu proposed probabilistic method for dams and seismic slope stability analysis based on Sarma method. T.S. Long et al. established a random seismic stability analysis method for embankment dam slope. Ai-Homoud et al. proposed a three-dimensional model of probabilistic seismic stability analysis of soil slope and embankment by use of the safety factor and critical displacement of slope failure and write a computer program PTDDSSA. W.H. Teng established a long-term reliability metrics for slope stability analysis based on the simplified Bishop method and considered the earthquake duration, the randomness of seismic intensity, and the variability of soil parameters.

Previous studies discussed slope stability probabilistic seismic analysis from different perspectives and methods. In general, research in this area has just begun, and there are a lot of issues that need further study.

1.2.5 Prediction Model for Earthquake-Induced Landslide Hazard

Prediction model for earthquake-induced landslide hazard range is used to predict landslide slip distance, which is a substantive issue related to disaster prevention and mitigation department and pressing concerns of local residents. However, because the formation mechanism of landslide motion mechanism, sliding speed, and other research is still in the exploratory stage, landslide hazard range prediction is facing challenges. Hazard range prediction has not yet been solved, research is relatively rare. Summary landslide sliding distance prediction researches can be divided into the following three categories:

(1) Empirical statistical prediction model
Such models established statistical formula-associated sliding distance based on the influential factors of landslide slipping distance and slope geometric features. Representative research has been undertaken by Scheidegger, who proposed statistical relationship between the landslide volume and the ratio of vertical height and horizontal fall distance. Tucson proposed statistical relationship among the ratio of vertical height and horizontal fall distance, tangent of the slope angle, and landslide volume. Wang Nian Qin and Zhang Zhuoyuan proposed a statistical model for predicting landslide sliding distance

of typical loess landslides. Li Xiuzhen et al. developed 5.12 earthquake-induced landslide sliding distance statistical model by using binary linear regression to horizontal sliding distance landslide, landslide volume, landslide edge height difference, and initial slope angle.

(2) Deterministic physical prediction model
These models are mainly based on the particle dynamics, combine the landslide movement patterns and the law of conservation of momentum. Representative results are sled model by Heim, geotechnical sled model, estimation the friction coefficient model and collision hypothesis, etc.

(3) Numerical simulation prediction model
These models predict the landslide sliding distance based on the finite element calculation method, discrete element, etc. numerical method, like Hungr et al., J.P. Qiao et al., D.H. Zhong et al.

Empirical statistical prediction model is more practicable than deterministic physical prediction model and numerical prediction model in the regional spatial prediction and is easy to establish the correlation between landslide characteristic parameters and sliding distance.

1.3 Problems

Comprehensive analysis of these findings shows further improvement and development are needed in shaking table simulation technology, dynamic characteristics of the slope under earthquake, deformation characteristics and slope instability mechanism, dynamic stability evaluation method of the slope under earthquake and earthquake-induced landslide hazard range prediction model.

(1) Research on dynamic characteristics of rock slope stays in situ observations, shaking table tests, and numerical simulation, few studies have been conducted on theoretical solution to seismic properties of rock slope. However, existing theoretical solutions to seismic properties of rock slope were derived from simple harmonic motions. The incident frequency and amplitude remain constant, but the actual seismic waves change frequency and amplitude gradually over time. This has resulted in the limited application of the above-mentioned theoretical solution.

(2) The most commonly used methods for numerical analysis are finite element method, finite difference method, and the traditional discrete element method. However, the finite element method and finite difference method is based on the continuous medium mechanics. These two methods are applicable to small-deformation cases. For rock slope large deformation dynamic analysis, they have some limitations, especially for rock slope containing numerous discontinuities. Traditional discrete element method is able to calculate the large deformation, but there are still some problems in the calculation of slope

stability; initial state is dependent on the distribution of structure surface, but structural surface parameters depend on the empirical value; furthermore, it cannot define the randomness and complexity of structure surface. Therefore, a new numerical method is needed. Finite element method will be adopted when the deformation is small, and traditional discrete element method while large deformation.

(3) As for the deformation characteristics and slope instability mechanism under earthquake, traditional research methods are mainly concentrated in time-domain and frequency-domain analysis method. Time-domain analysis can only consider temporal aspects of the seismic response, such as displacement, velocity, and acceleration. Frequency-domain analysis only considers the Fourier spectrum and the energy spectrum, etc. Seismic wave is a complex non-stationary signal; its frequency and amplitude characteristics gradually change over time, its time, frequency, and amplitude is a large coupled three-dimensional system. Therefore, only time-domain or frequency-domain analysis will ignore some significant information on the deformation characteristics and slope instability mechanism under earthquake, causing an incomplete understanding to this issue.

(4) In terms of seismic stability of slopes, probability analysis is of a late start and not mature so far; numerical methods and test methods have operational problems, high cost, and time-consuming. The three methods are difficult to be promoted. Regardless of the widely use of slice method, it only takes the domain characteristics of ground motions into consideration, not fully considers the characteristics of ground motion, less accurate. Quasi-static method is currently the most widely used method, but it cannot considered earthquake frequency and duration characteristics, since the earthquake usually be described by amplitude, frequency, and duration, this greatly reduces identification accuracy of the pseudo-static method.

(5) Currently, most widely used seismic landslide hazard range prediction model considers the landslide volume (V) and maximum vertical sliding distance of landslide (H). However, a large number of research results show peak ground acceleration, slope angle and so on have a high correlation to landslide distribution and scale, which greatly reduces the current prediction accuracy of landslide prediction model.

1.4 Research Objectives and Content

Slope stability analysis is one of the oldest geotechnical engineering problems, a wealth of experiences have been accumulated so far. Static stability analysis slope has matured; dynamic stability analysis of slope involves multiple disciplines, such as engineering geology, soil mechanics, geophysics, and earthquake engineering. This makes the seismic stability analysis of slope be of great difficulty. So far, there has not been a mature theory and standards to follow, which can be reflected by the relevant provisions of the seismic code. Therefore, the study of seismic slope stability analysis is all the more urgent.

Based on this, this project will aim to questions raised in "**1.3 existing problems**" section, Wenchuan earthquake as the background, from the macro- and micro-levels, dynamic characteristics of slope as the main line, in-site investigation, numerical analysis, shaking table test, and theoretical analysis as a mean, to explore the dynamic characteristics of seismic rock slope. On this basis, cover-type slope is chosen as the research project, in-depth analysis to the slump mechanism, proposes a system that can consider three factors (amplitude, frequency, and duration) and predicts landslide hazard range. Specific contents are as follows:

1. Dynamic characteristics and influencing factors rock slope

Bedrock and overburden slope is a special form of rock slope; understanding the dynamic characteristics of rock slope under earthquake is to key reveal the failure mechanism of bedrock-debris slope under earthquake. Data from seismic monitoring array, large-scale shaking table test, and numerical simulation will be used; the effects of slope angle, seismic intensity, incident wave type, slope type, and local slope form will be explored. Then, the theoretical solution for double-sided rock slope under earthquake will be derived based on the elastic wave theory. Dynamic characteristics of earthquakes rock slope will be discussed quantitatively.

2. Deformation characteristics and sliding mechanism of bedrock-debris slope under earthquake loading

Deformation characteristics and sliding mechanism of bedrock-debris slope under earthquake loading are premised for slope stability evaluation, disaster prevention, and mitigation. In this section, based on the research on the dynamic characteristics of rock slope, chosen bedrock-debris slope as research object, summarize the geological model, utilizing novel time–frequency–amplitude analysis techniques, combining with the data from shaking table test and numerical simulation, to explore the deformation characteristics and sliding mechanism of bedrock-debris slope under earthquake loading form both time and frequency domains.

3. Time–frequency analysis method for seismic stability of bedrock-debris slope

With the depth understanding of dynamic stability of slope, the shortcoming of pseudo-static method to analyze the stability of slope has been recognized. Time–frequency-domain analysis has become a trend for slope stability analysis. Based on elastic wave theory, geological model will be summarized, Hilbert–Huang transformation will be employed to propose a time–frequency-domain analysis method for slope seismic stability analysis, and the method developed in this project will be verified by comparing with the results of shaking table tests and numerical simulations.

4. Prediction model for disaster range of bedrock-debris landslides triggered by earthquake

The effective method to deal with large-scale, widespread earthquake-induced landslides is to consider this issue even in the planing stage. In other words, relevant calculation parameters should be determined after seismic stability assessment, and then, the evaluation to the landslide disaster should be conducted by employing disaster range prediction model. Most of the disaster range prediction models only take sliding volume and vertical sliding distance into consideration. However, a large number of studies shown that peak acceleration and slope angle have a strong correlation to distribution and scale of landslides. Therefore, a prediction model will be proposed in this section based on the field investigation results and peak ground acceleration attenuation model, full consideration will be paid to the effect of mountain height, slope angle, peak ground acceleration, slump volume, etc.

Chapter 2
Seismic Array Monitoring Results Analysis

Bedrock and overburden slope is a special kind of rock slope. In order to reveal the failure mechanism of it under earthquake, its seismic dynamic response must be explored first. The dynamic response of rock slope includes acceleration, velocity, displacement, stress, and strain. The research results show that the inertia force caused by acceleration is the main reason of slope deformation and failure, which could provide reference to the seismic design of slope engineering. Hence, acceleration and its distribution rules are basic materials to evaluate the seismic dynamic characteristics of slope engineering.

So far, the research method for seismic ground motion characteristics of slopes mainly include field investigation, model tests, numerical simulation, and theoretical analysis, while seismic array monitoring is an important measure which can most reliably and truly reflect the seismic ground motion characteristics of slopes, whose monitoring data could provide valuable original data for seismic response analysis. Thus, this section will adopt the monitoring data of seismic array of rock slope in the Sishan park, Zigong in "5.12 Wenchuan Earthquake," explain the real motion characteristics of slopes under seismic effects from the macro perspective, and visually analyze the seismic ground motion characteristics of rock slope so as to provide basic reference for following shaking table test, numerical simulation, and theoretical derivation of rock slopes as well as the formation mechanism of bedrock and overburden slopes.

2.1 General Condition

Zigong City locates in southern Sichuan Basin, and its distance to the epicenter of Wenchuan earthquake is about 227 km, it is composed of hills and valleys. Elevation decreases from northwest to southeast, mainly in the range of 250–450 m, as illustrated in Fig. 2.1. The elevation of downtown area is from 280 to 400 m, ground relative elevation is small, as shown in Fig. 2.2. Seismic array

Y. Changwei et al., *Slope Earthquake Stability*,
DOI 10.1007/978-981-10-2380-4_2

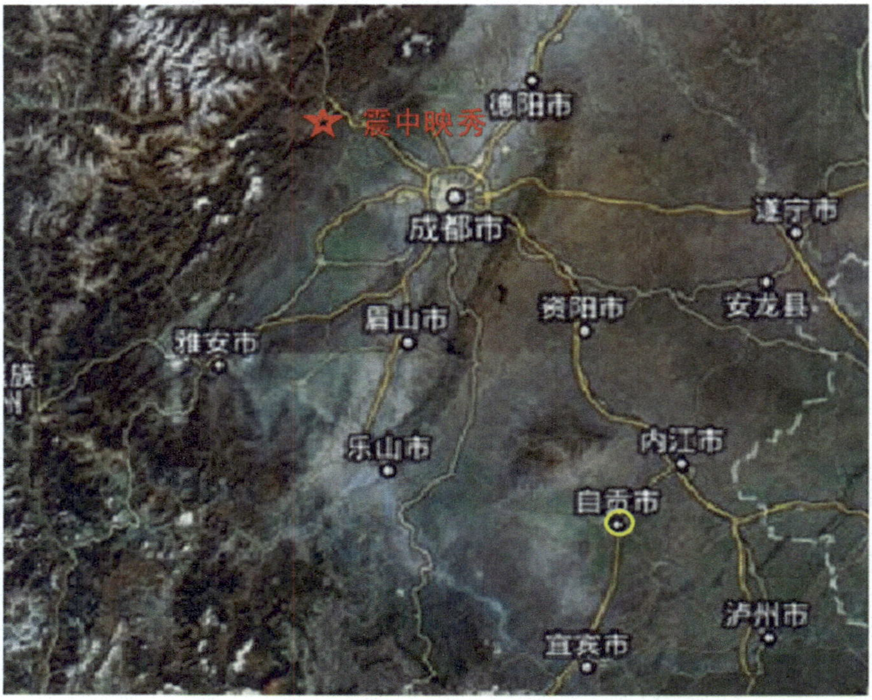

Fig. 2.1 Location of Zigong City

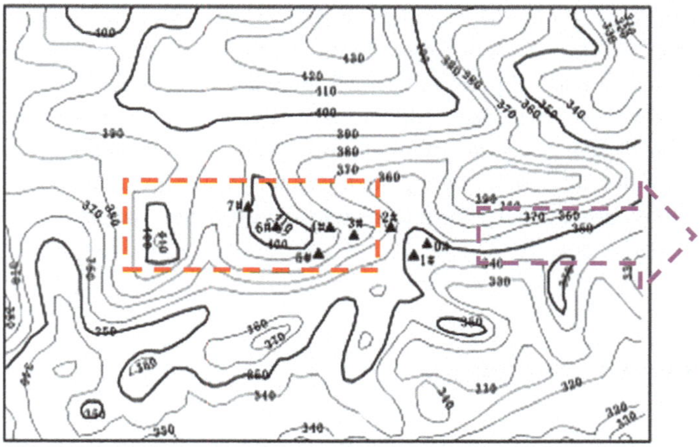

Fig. 2.2 Plane distribution map of Zigong topography monitoring points

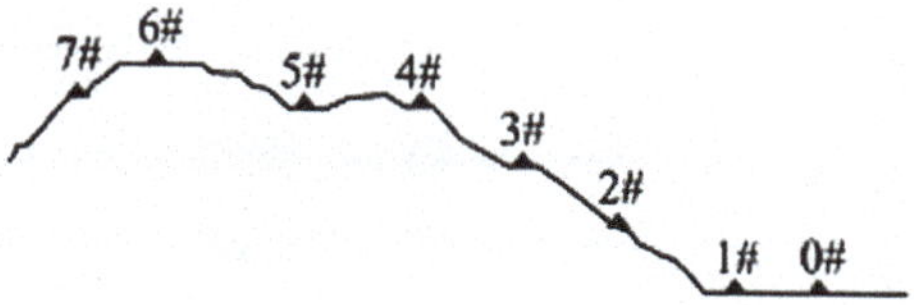

Fig. 2.3 Distribution map of seismic array monitoring points

Table 2.1 Table of array parameters of Zigong topography

Station name	Elevation	Horizontal distance to reference station	Rock property
0#	345.0	51.55	Soil layer
1#	345.0	0.00	Jurassic bedrock
2#	367.0	41.64	Jurassic bedrock
3#	385.0	116.82	Jurassic bedrock
4#	392.0	171.41	Jurassic bedrock
5#	390.0	258.98	Jurassic bedrock
6#	417.0	304.55	Jurassic bedrock
7#	397.0	345.72	Jurassic bedrock

locates on a hill in Western Hills Park. The array consists of eight stations (0#–7#). Each station is equipped with digital seismograph with ETNA recorder and ES-T-type accelerometer. The maximum height difference is about 72 m between seismograph (6#) in the top of the hill and seismograph (0#, 1#) in the bottom of the hill. 0# station is located on soil site, 1# station is located on rock site, and other stations are located on rock site with different elevations (Fig. 2.3 and Table 2.1).

2.2 Monitoring Data of Seismic Array in 5.12 Wenchuan Earthquake

Wenchuan earthquake (Ms 8.0) occurred in China on May 12, 2008, and the seismogenic fault is central Longmen Shan rupture, with 230 km of surface rupture. Terrain is extremely complex, the elevation of upward plate is about 2000–3000 m, the elevation of the Chengdu Plain on the downward plate is just 500 m. The complex terrain of Longmen Shan fault zone exerts a significant effect on the Wenchuan earthquake. According to information provided by Chinese seismic array network, the seismic station in Zigong city has monitored relatively complete seismic data, which laid a solid foundation for the seismic response analysis of slope. The monitoring results are displayed in Figs. 2.4, 2.5, and 2.6.

Table 2.2 shows the peak ground accelerations of monitoring points in Xishan Park. To study the variation of PGA along elevation, so PGA in EW, NS, and UD directions are extracted, respectively, as shown in Table 2.2.

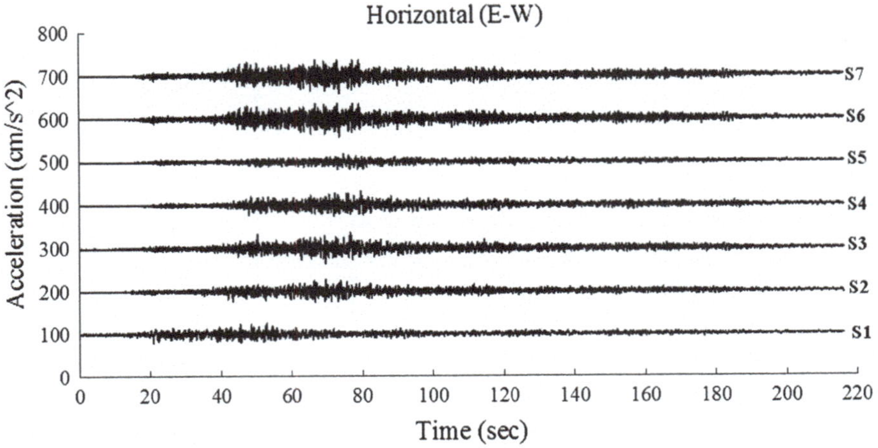

Fig. 2.4 The measured seismic ground motion time history curve in EW direction of S1–S7

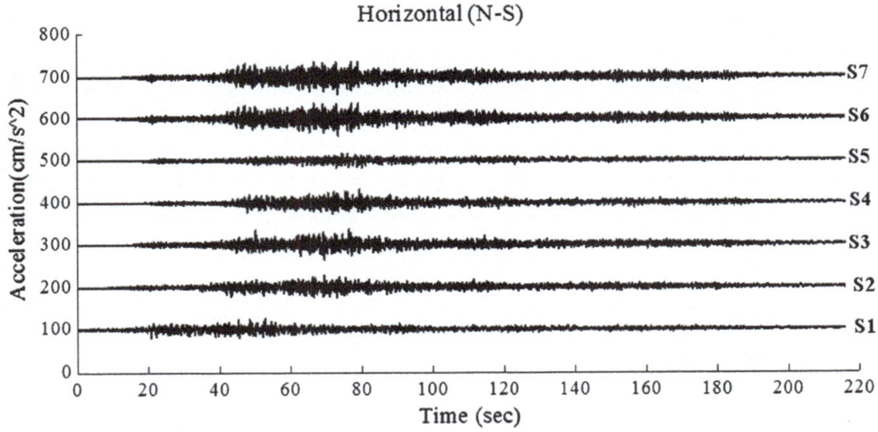

Fig. 2.5 The measured seismic ground motion time history curve in NS direction of S1–S7

2.3 Response Characteristics Analysis

According to the data from Zigong seismic array, elevation amplification effect will be discussed in this section. The #1 is chosen as reference point, PGA amplification factor is defined as the ratio of PGA at any point to that of #1, as shown in Figs. 2.7, 2.8, and 2.9. It should be pointed out that the elevation of each site is the relative elevation to #1.

From Figs. 2.7, 2.8, and 2.9, the PGA amplification factor increases with the relative elevation in three directions. The maximum value of PGA amplification factor in EW direction is 1.768 at #1, that in ES direction is 1.717 at #7, and that in

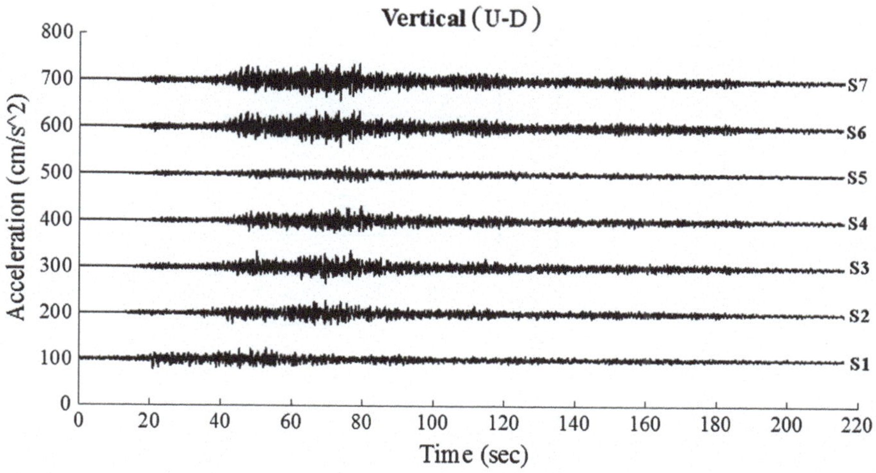

Fig. 2.6 The measured seismic ground motion time history curve in UD direction of S1–S7. *Note* all the time history curves are drew by uniform scale, and the stations are numbered S1–S7

Table 2.2 PGA response characteristics of Zigong topography monitoring array (cm/s/s)

No.	Relative elevation (m)	PGA (cm/s^2)			Acceleration amplification coefficient		
		EW component	NS component	UD component	EW component	NS component	UD component
1#	0	23.036	26.489	14.726	1	1	1
2#	22	28.078	29.952	15.953	1.219	1.131	1.083
3#	40	34.863	32.533	18.336	1.513	1.228	1.245
4#	47	32.531	32.226	19.719	1.412	1.217	1.339
5#	45	33.030	42.169	17.240	1.434	1.592	1.171
6#	72	40.735	42.244	19.742	1.768	1.595	1.341
7#	52	39.299	45.489	16.049	1.706	1.717	1.090

Fig. 2.7 Acceleration elevation amplification effects in EW direction

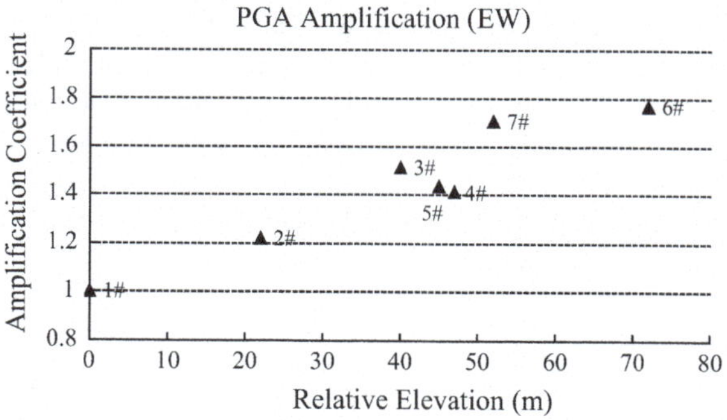

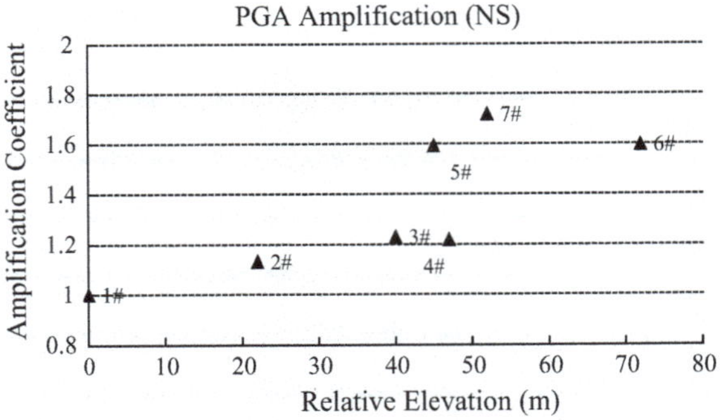

Fig. 2.8 Acceleration elevation amplification effects in NS direction

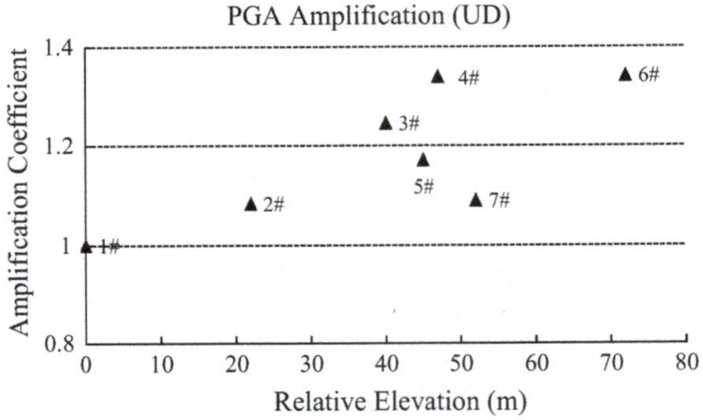

Fig. 2.9 Acceleration elevation amplification effects in UD direction

UD direction is 1.341 #6. The values of horizontal directions are very close, both are great than that in vertical direction. This may be caused by the fact that EW is close to the direction of free surface and the NS is close to the tendency of the slope. The free surface amplification effect results in a more intensive amplification in EW direction than NS direction. It is noteworthy that the PGA amplification factor of #7 is smaller that of #5 while the relative elevation of the former site is larger than the later site. This can be contributed to the fact that the #5 locates in a local canyon. The reflection, refraction, and transmission effect will be triggered while the up propagation seismic waves encounter the interface between the soil layer and underlying bedrock, which lead to a much more intense ground motion at #5. This indicates that the acceleration effect is dependent on direction, namely the direction corresponding to the free surface has the largest amplification value, the vertical direction has the smallest amplification, and the direction parallel to the slope tendency is moderate.

2.4 Frequency Spectrum Response Characteristics

Based on the monitoring results of seismic dynamic response of seismic array of Zigong topography, this section analyzes the frequency spectrum response characteristics of rock slopes from both Fourier spectrum and response spectrum.

2.4.1 Fourier Spectrum Along Elevation

Fourier spectrum is the method to change complex seismic acceleration time history α(t) into N frequency combinations by discrete Fourier transform technique:

$$\alpha(t) = \sum_{i=1}^{N} A_i(\omega)\sin[\omega_i t + \varphi_i(\omega)] \tag{2.1}$$

In the formula, $A_i(\omega)$, $\varphi_i(\omega)$ are the amplitude and phase angle of vibration component of angular frequency ω_i;

Formula (2.1) can be rewritten into:

$$\alpha(t) = \sum_{i=1}^{N} A_i(i\omega)\mathrm{e}^{\mathrm{i}\omega_i t} \tag{2.2}$$

In the formula, $i = \sqrt{-1}$, complex function $A(i\omega)$ is Fourier spectrum, and its module $|A(i\omega)|$ is the amplitude spectrum.

Calculate the Fourier spectrum in EW, NS, and UD direction from S1 to S7 based on the monitoring results of Zigong topography seismic array, and carry out smooth effect to the Fourier spectrum, which are shown in Figs. 2.10, 2.11, 2.12, 2.13, 2.14, 2.15, and 2.16.

Fig. 2.10 Fourier spectrum of three-direction acceleration of S1

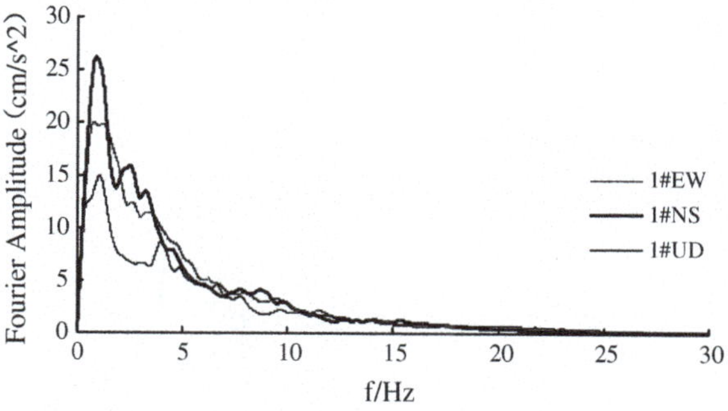

Fig. 2.11 Fourier spectrum of three-direction acceleration of S2

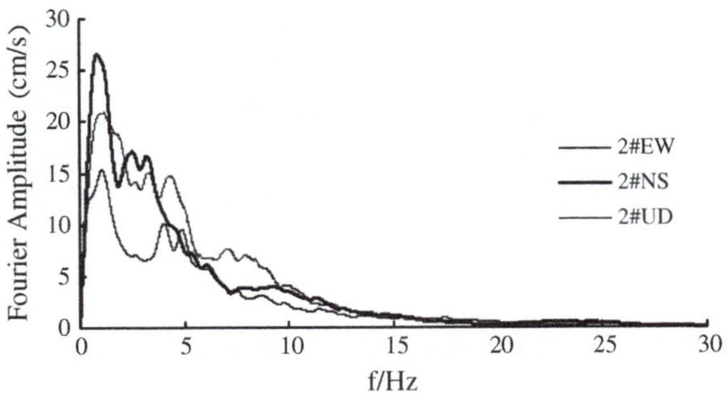

Fig. 2.12 Fourier spectrum of three-direction acceleration of S3

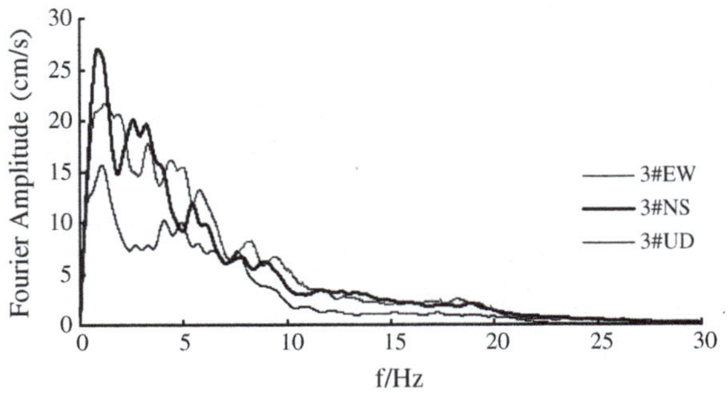

Fig. 2.13 Fourier spectrum of three-direction acceleration of S4

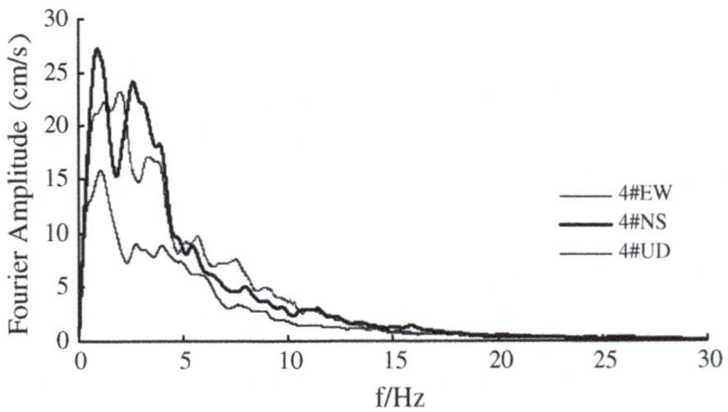

Fig. 2.14 Fourier spectrum of three-direction acceleration of S5

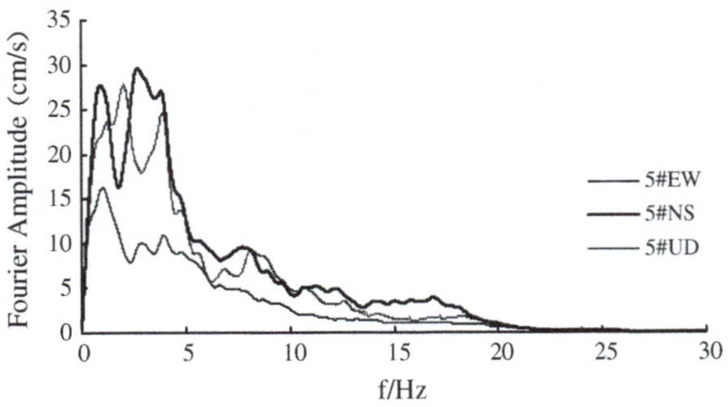

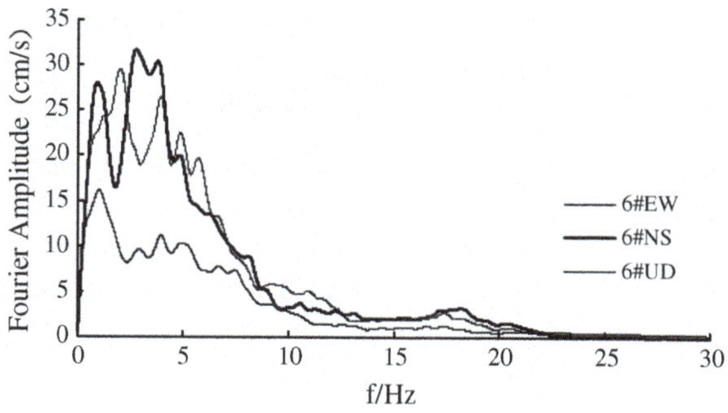

Fig. 2.15 Fourier spectrum of three-direction acceleration of S6

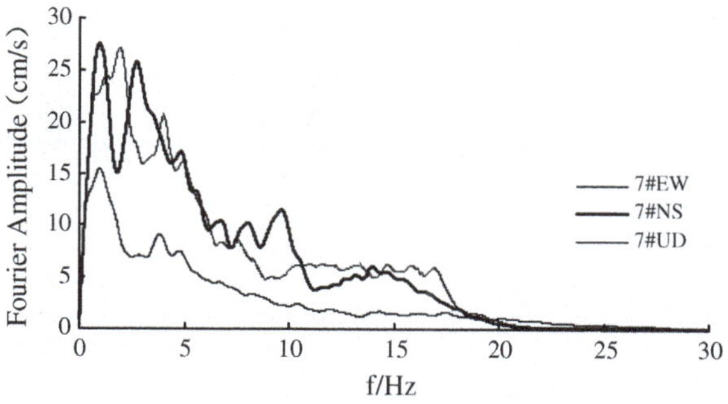

Fig. 2.16 Fourier spectrum of three-direction acceleration of S7

Figures 2.10, 2.11, 2.12, 2.13, 2.14, 2.15, and 2.16, show that as for the acceleration Fourier spectrum in EW direction, the dominant frequency of 1#–4# is in the range of 1–2 Hz, these of 5# and 6# present a double peak pattern, dominant frequency is in the range of 1–2 Hz and 4–5 Hz, its amplitudes increase with the elevation. In terms of the acceleration Fourier spectrum in ES direction, the dominant frequency of 1# and 2# is in the range of 1–2 Hz, these of 3#–6# present a double peak pattern, and dominant frequency is in the range of 1–2 Hz and 3–5 Hz, its amplitudes increase with the elevation. In UD direction, the dominant frequency is in the range of 1–2 Hz, its amplitude stays unchanged and is independent of elevation. According to the evaluation method to natural frequency proposed by J.Y. Xu et al. the natural frequency of this slope is 4.02 Hz, hence, the components within 3.0–5.0 Hz are amplified.

Generally, dominant frequency band of horizontal acceleration is in the range of 1.0–2.0 Hz and 3.0–5.0 Hz. The frequency components are close to the natural frequency of the slope 3.0–5.0 Hz, so acceleration amplifies dramatically, and the Fourier amplitude spectrum transform form single mode to double peak pattern. The dominant frequency band of vertical acceleration is 1.0–2.0 Hz, which is independent of elevation.

2.4.2 Seismic Ground Motion Response Spectrum Along Elevation

In essence, seismic response spectrum is the characteristics of seismic intensity and spectrum. The response spectrum is composed of maximum responses of numerous single degree of freedom systems. The response spectra are illustrated in Ging Figs. 2.17, 2.18, 2.19, 2.20, 2.21, and 2.22 with damping ratio 5 %.

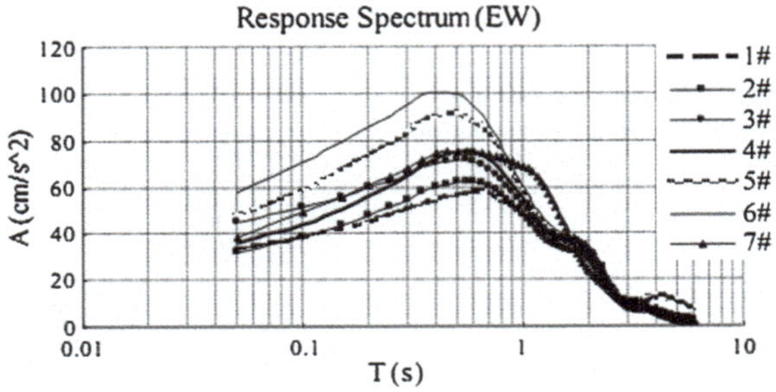

Fig. 2.17 Acceleration response spectrum of all stations in EW direction

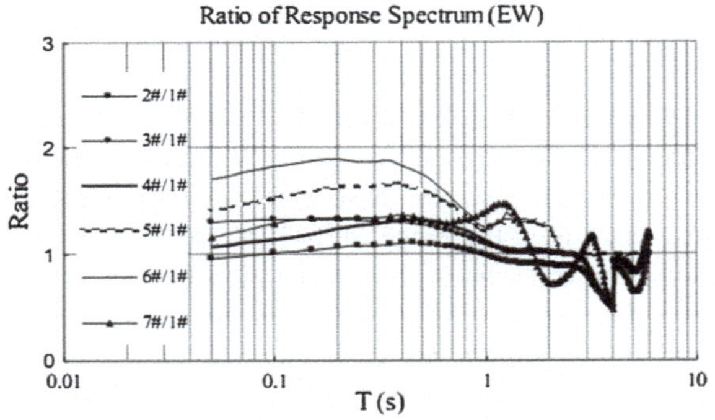

Fig. 2.18 Acceleration response spectrum of all stations in EW direction

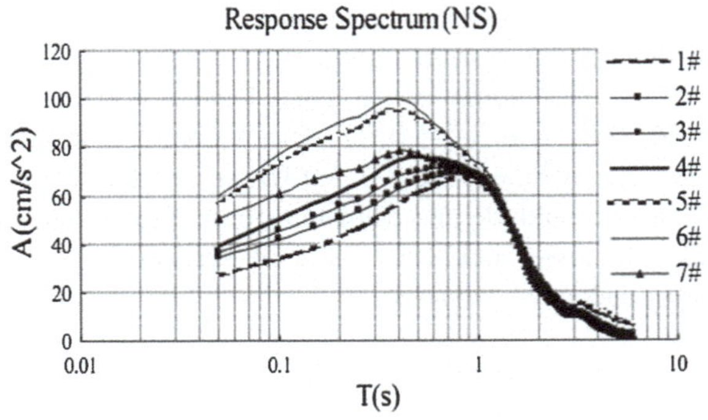

Fig. 2.19 Acceleration response spectrum of all stations in NS direction

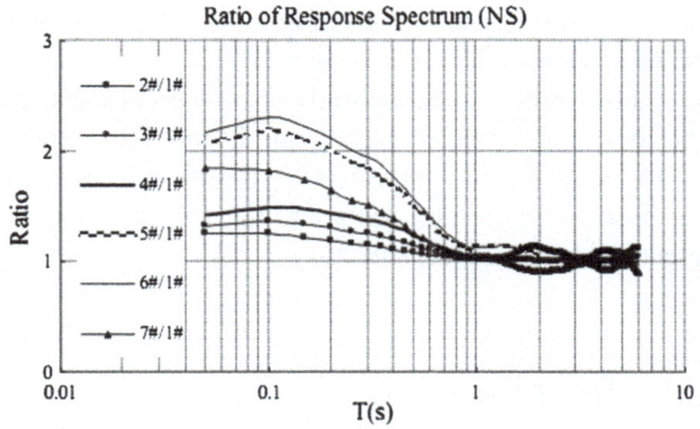

Fig. 2.20 Acceleration response spectrum of all stations in NS direction

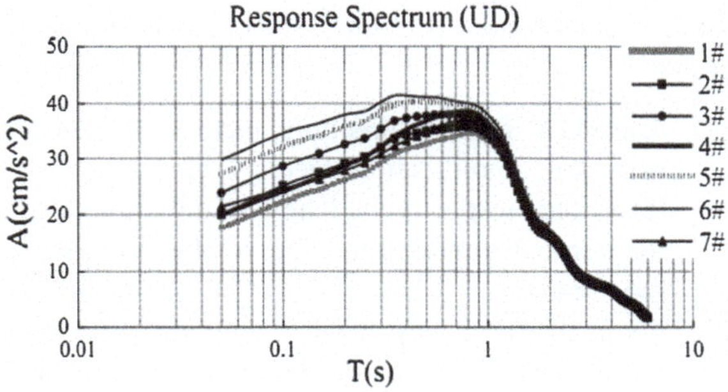

Fig. 2.21 Acceleration response spectrum of all stations in UD direction

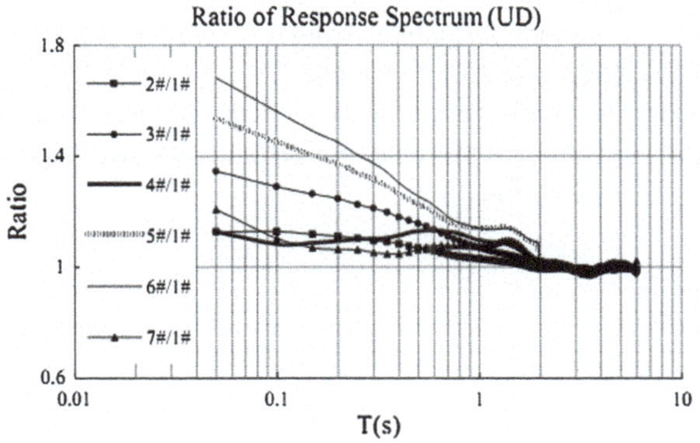

Fig. 2.22 Acceleration response spectrum of all stations in UD direction

Figures 2.17, 2.18, 2.19, 2.20, 2.21, and 2.22 show that the dominant periods of response spectrum in both EW and NS directions are in the range of 0.3–0.7 s, and the amplitudes increase with the elevation for $T \leq 1$ s, while $T > 1$ s, the response spectrum of each point are almost the same. In terms of the vertical direction, the dominant periods of response spectra are in the range of 0.6–1.0, and the amplitudes increase with the elevation for $T \leq 1$ s, while $T > 1$ s, the response spectrum of each point are almost the same, the amplitude are about one-third to half of that of horizontal direction, and the amplitudes of 2#, 4#, and 7# have little difference. It is noteworthy that the 5# is located in a local canyon. Even though the elevation of 5# is smaller that 7#, the amplitude of 5# is greater than that of 7#, which indicates that the local site condition has a significant effect on ground amplification.

In general, the amplification coefficient of acceleration response spectrum increases with the elevation, and the amplification value in horizontal direction is greater that in vertical direction.

2.5 Brief Summary

Through the analysis of the seismic array of Xishan Park, its PGA, Fourier spectrum, and response spectrum are obtained, and the following conclusions can be drawn:

(1) The PGA in three directions increase nonlinearly with the increase of elevation, the amplification factor is 1–1.768, 1–1.717, and 1–1.341 for three directions, respectively. The EW direction has the largest amplification value, the vertical direction has the smallest amplification, and the NS direction is moderate.
(2) dominant frequency band of horizontal acceleration is 1.0–2.0 Hz and 3.0–5.0 Hz, these frequency components close to the natural frequency of the slope 3.0–5.0 Hz are amplified dramatically, and the Fourier amplitude spectrum transform form single peak mode to double peak pattern. The dominant frequency band of vertical acceleration is 1.0–2.0 Hz, which is independent of elevation.
(3) The amplification factor of acceleration response spectrum increases with the elevation, and the amplification value in horizontal direction is greater that in vertical direction.

Chapter 3
Shaking Table Test of Rock Slopes

As it is known to all that the most direct and reliable way to study the response characteristics of rock slope seismic action is to conduct field test through seismic array network, which could intuitively reflect the motion features and seismic dynamic characteristics of rock slopes so as to provide precious original data for the future work. But because of the destructiveness, occasionality of earthquake, and the uncertainty of many other factors, the field test consumes large amount of time, labor, and cost, and it is not repetitive as it can only study on single rock slope. By contrast, the similar model test in the laboratory could effectively compensate the shortcomings of field tests, which is the model test technique, based on the similarity theory, using the likeness or similarity between or among things or phenomena to study natural laws. The structure models are small in size, easy to make, assemble and disassemble—which save materials, time, and labor—and have high pertinency and accurate and reliable data collection. In addition, one model could be used for tests under different working conditions. It is especially applicable for research areas in which data results are difficult to obtain through theoretical analysis and field test measures and is quite effective for analysis and comparison of theoretical research results. Therefore, with macro-analysis of rock slopes using seismic array network monitoring data, conducting the similar model test in the laboratory could enrich research results and its contents.

Currently, the shaking table test is the major means to simulate earthquake in laboratory, which has similar ground movements and impacts on structures as real earthquake occurs. It is the important way and a measure to study failure mechanism and failure model of earthquakes and to evaluate the global seismic capacity, which has been widely applied to the theoretical research and practices of earthquake engineering. Thus, this chapter will begin by deducing the similarity law of rock slope shaking table test, and then, using the test results, study the response rules of rock slope acceleration under earthquake from the domain of both time and frequency, and the influence on rock slope dynamic characteristics of bevel angle, seismic intensity, ground motion input type, etc., so as to lay a foundation for the study on instability mechanism of bedrock and overburden layer slopes under earthquakes.

Y. Changwei et al., *Slope Earthquake Stability*,
DOI 10.1007/978-981-10-2380-4_3

3.1 Large-Scale Shaking Table Test Design

3.1.1 Purpose

In order to study the dynamic response rules of homogeneous rock slopes, this paper, taking the 5.12 Wenchuan earthquake as background, and relying on Sichuan-Tibet highway environmental disaster evolution and the project risk, the major project of NSFC, carries out large-scale shaking table physical simulation tests. During the experimental process, tracking the seismic dynamic response of acceleration and displacement, etc., of different position and elevation of rock slopes, revealing the response rules of rock slope acceleration under earthquake, and the influence of bevel angle, seismic intensity, seismic wave input type, and input wave frequency on the rock slope dynamic characteristics, which provides theoretical supports for further study on deformation and failure mechanism of bedrock and overburden slopes, and for reasonable explanation for Wenchuan earthquake inducing secondary geological disasters.

3.1.2 Installation

The shaking table test installation adopts the large-scale shaking table of three axles and six degrees of freedom designed by the Seismic Laboratory of the second institute of Nuclear Power Institute of China. The installation is currently the earthquake shaking table with the largest capacity and the most advanced performance in China. The surface size is 6 m × 6 m, the maximum load is 60 t, and the frequency range is 0.1–80 Hz. The shaking table could load from three axles and six degrees of freedom simultaneously, and when in full load, the maximum displacement in X and Y axles is ±150 mm with the maximum acceleration of ±1.0 g; the maximum displacement in Z axle is ±100 mm with the maximum acceleration of ±0.8 g.

In the shaking table test of geotechnical engineering, the typical model containers include laminar shear model box, dish-type container, ordinary rigid container plus lining, and flexible container. However, given the requirements of model scale and rigidity, this test adopts the rigid model container made of steel plate and steel-reinforced concrete, and its inner-space size is 5.0 m × 5.0 m × 2.1 m (length × width × height). The baseboard and framework of the model container are made of steel plate, angle steel, and channel steel, and the total weight of the model container is 7473.55 kg. See Fig. 3.1.

3.1.3 Similarity System

For the shaking table test of rock slope model, it is crucial to keep similarity between the model and the prototype. While due to difficulties in test materials and

Fig. 3.1 Model containers of shaking table test

technologies, most shaking table tests could only adopt scale model for experiments, and because of the intricate background, complex terrain, and large scale of Wenchuan earthquake-induced landslides and collapses, it is hard to make the model completely consistent with the prototype. Therefore, for the main purpose of this shaking model test, we could only make one or some indicators similar to or proximately similar to that of the prototype for the experiment.

There are mainly three kinds of similarity relations in the physical simulation test: geometric similarity, dynamic similarity, and kinetic similarity. If two physical phenomena meet the similarity conditions in geometric similarity, dynamic similarity, and kinetic similarity, then they are considered as similar phenomena. Among these three kinds of similarity relations, geometric similarity is easy to achieve, while dynamic similarity is controlled by and shown with geometric similarity and kinetic similarity. Thus, dynamic similarity is the key among the three.

3.1.3.1 Three Theorems of Similarity

Physical phenomenon is only similar on certain conditions; the results of similarity simulation test could be applied to the prototype only under certain conditions. The three theorems of similarity are the summary of such conditions.

(1) The first theorem of similarity

The first theorem of similarity is defined as the similarity rules of the similar phenomena are equal, the similarity index is equal to 1, and the single-valued conditions are similar.

The single-valued condition refers to features in which individual phenomena differ from the same phenomena, which includes: geometric condition, physical condition, boundary condition, and initial condition. Geometric condition refers to the shape and size of objects in the process, physical condition refers to the physical property of objects, boundary condition refers to the external constraints on the objects' surface, and initial condition refers to certain features of the objects at the initial time. Taking research on heat conduction process for example, the shape and geometric size of the object are the geometric conditions, the specific heat capacity and the heat conductivity coefficient are the physical conditions, the heat conductivity coefficient of the surface medium belongs to boundary condition, and the temperature of the object at the initial time is the initial condition.

(2) The second theorem of similarity

The second theorem of similarity is also called "π theorem," which could be defined as follows: If the phenomena are similar, then the parameter relations used to describe them could be converted into functional relations of similarity rules, and the function equation of the similarity rules is the same.

As the similarity rules are dimensionless, therefore, the physical equation used to describe similar phenomenon could be converted into dimensionless equation of similarity rules.

$$f(a_1, a_2, \ldots, a_k, a_{k+1}, a_{k+2}, \ldots, a_n) = 0 \tag{3.1}$$

$$F(\pi_1, \pi_2, \ldots, \pi_{n-k}) = 0 \tag{3.2}$$

In Eq. (3.1), a_1, a_2, …, a_k are basic quantities, and a_{k+1}, a_{k+2},…, a_n are derived quantities, which shows there are $(n - k)$ similarity rules.

The second theorem of similarity provides theoretical support for the spread of model test results. This is because if the two phenomena are similar, according to the second theorem of similarity, the model test results could be spread into the prototype and bring perfect explanation for it. The first and second theorems of similarity state the common features of similar phenomena and provide support for the spread of model test results.

(3) The third theorem of similarity

The third theorem of similarity could be defined as if two phenomena can represent by the same function equation and are similar in single-valued condition, and at the same time, the similarity rules composed of single-value condition are equal, then the two phenomena are similar. In the engineering practice, it is quite difficult to make model and prototype completely meet the requirements of the third theorem of similarity, and it is even impossible. In situation, we can, according to the features of the objects, reasonably select the high-impact factors, seize the main contradiction of the phenomena, and omit the secondary factors to achieve the model test, which is the so-called approximate simulation. The success of approximate simulation lies in reasonable selection of influence factors. Although it cannot make sure all the similar conditions are satisfied, it ensures the similarity of the main factors, so the accuracy of the research results generally can meet the requirement of engineering practice.

3.1.3.2 Selection and Adjustment of Similarity Rules

When carrying out model test similarity design, we could adopt similarity transformation method, dimensional analysis method, and matrix method to work out the similarity rule, i.e., the original rule. Ordinarily, the original rule needs process through transformation and adjustment on form, which is mainly because when different methods are adopted to derive similarity rules, the form of these rules might be different. Even for some single derivation method, position arrangement changes of certain parameters might lead to form changes of these rules.

It is difficult or even impossible to make model test meet all the similarity conditions when carrying out similarity tests. At this time, we should seize the main factors that influence the inherent law of phenomenon, i.e., to satisfy the major rules and omit some secondary rules, which is a common measure used by approximate simulation. While it must be cautious to omit similarity rules to ensure, they indeed have little influence to the change laws of phenomena, which need in-depth analysis to the researching phenomena.

3.1.3.3 Determination of Related Physical Parameters

This large-scale shaking table test of rock slope relates to many physical parameters. There are 17 separate physical parameters after analysis and classification as follows: geometric dimensions L, gravitational acceleration g ($C_g = 1$), cohesive force c, dynamic modulus E, internal friction angle φ, dynamic Poisson's ratio μ, unit weight γ, shear wave velocity V_s, input acceleration A, duration T_d, frequency ω, angular displacement θ, linear displacement s (should guarantee $C_s = C_L$), response speed V, response acceleration a, stress σ, and strain ε.

3.1.3.4 Derivation of Similarity Criteria Through Matrix Method

The above 17 physical parameters need meet the physical equation (see Eq. 3.3). Then we adopt [M], [L], [T] as fundamental dimension, and adapt it to similarity rule equation of dimensionless parameters (see Eq. 3.4). Finally, the general expression of similarity rule is worked out (see Eq. 3.5).

$$f(L, c, \varphi, \gamma, E, \mu, v_{\mathrm{s}}, g, A, T_{\mathrm{d}}, \omega, s, \theta, \varepsilon, \sigma, V, a) = 0, \tag{3.3}$$

$$F(\pi_1, \pi_2, \ldots, \pi_{14}) = 0 \tag{3.4}$$

$$\pi_i = [L]^{a_1} \cdot [c]^{a_2} \cdot [\varphi]^{a_3} \cdot [\gamma]^{a_4} \cdot [E]^{a_5} \cdot [\mu]^{a_6} \cdot [v_{\mathrm{s}}]^{a_7} \cdot [g]^{a_8} \cdot [A]^{a_9} \cdot [T_{\mathrm{d}}]^{a_{10}} \cdot [\omega]^{a_{11}} \cdot [s]^{a_{12}} \cdot [\theta]^{a_{13}} \cdot [\varepsilon]^{a_{14}} \cdot [\sigma]^{a_{15}} \cdot [V]^{a_{16}} \cdot [a]^{a_{17}} \tag{3.5}$$

In the similarity rule, the dimension of the above 17 physical parameters is in Table 3.1.

Bringing the above dimension of main physical parameters into the general expression of similarity rule (3.6), we get:

$$M^0L^0T^0 = L^{a_1} \cdot (ML^{-1}T^{-2})^{a_2} \cdot (1)^{a_3} \cdot (ML^{-2}T^{-2})^{a_4} \cdot (ML^{-1}T^{-2})^{a_5}(1)^{a_6} \cdot (LT^{-1})^{a_7} \cdot (LT^{-2})^{a_8} \cdot (LT^{-2})^{a_9} \cdot (T)^{a_{10}} \cdot (\omega)^{a_{11}} \cdot (L)^{a_{12}} \cdot (1)^{a_{13}} \cdot (1)^{a_{14}} \cdot (ML^{-1}T^{-2})^{a_{15}} \cdot (LT^{-1})^{a_{16}} \cdot (LT^{-2})^{a_{17}} \tag{3.6}$$

Merging the same dimensions, we can get:

$$M^0LT^0 = M^{a_2+a_4+a_5+a_{15}} \cdot L^{a_1-a_2-2\cdot a_4-a_5+a_7+a_8+a_9+a_{12}-a_{15}+a_{16}+a_{17}} \cdot T^{-2a_2-2\cdot a_4-2\cdot a_5-a_7-2\cdot a_8-2a_9+a_{10}-a_{11}-2\cdot a_{15}-a_{16}-2\cdot a_{17}} \tag{3.7}$$

According to the uniformity of dimensions, by Eq. (3.8):

$$\begin{cases} a_2 + a_4 + a_5 + a_{15} = 0 \\ a_1 - a_2 - 2\cdot a_4 - a_5 + a_7 + a_8 + a_9 + a_{12} - a_{15} + a_{16} + a_{17} = 0 \\ -2a_2 - 2a_4 - 2a_5 - a_7 - 2a_8 - 2a_9 + a_{10} - a_{11} - 2a_{15} - 2a_{16} - 2a_{17} = 0 \\ a_3 + a_6 + a_{13} + a_{14} = \text{any number} \end{cases} \tag{3.8}$$

Use matrix method to solve the 14 derived similar constants. Table 3.2 is using matrix method to derive similarity criteria.

Table 3.1 Dimension of main physical parameters

Physical parameters	Quality system	Physical parameters	Quality system
$[L]$	L	$[A]$	LT^{-2}
$[c]$	$ML^{-1}T^{-2}$	$[T_{\mathrm{d}}]$	T
$[\varphi]$	1	$[\omega]$	T^{-1}
$[\Upsilon]$	$ML^{-2}T^{-2}$	$[s]$	L
$[E]$	$ML^{-1}T^{-2}$	$[\theta]$	1
$[\mu]$	1	$[\varepsilon]$	1
$[V_{\mathrm{s}}]$	LT^{-1}	$[\sigma]$	$ML^{-1}T^{-2}$
$[g]$	LT^{-2}	$[V]$	LT^{-1}
$[\alpha]$	LT^{-2}		

Table 3.2 Similarity criteria derived from matrix method

物理量	目标量																	解得组合关系式
	c	φ	E	μ	V_s	g	A	T_d	s	θ	ε	σ	V	a	L	γ	ω	
	a_2	a_3	a_5	a_6	a_7	a_8	a_9	a_{10}	a_{12}	a_{13}	a_{14}	a_{15}	a_{16}	a_{17}	a_1	a_4	a_{11}	
π_1	1	0	0	0	0	0	0	0	0	0	0	0	0	0	-1	-1	0	$\pi_1 = \frac{c}{L\gamma}$
π_2		1	0	0	0	0	0	0	0	0	0	0	0	0	0	0	0	$\pi_2 = \varphi$
π_3			1	0	0	0	0	0	0	0	0	0	0	0	-1	-1	0	$\pi_2 = \frac{E}{L\gamma}$
π_4				1	0	0	0	0	0	0	0	0	0	0	0	0	0	$\pi_4 = \mu$
π_5					1	0	0	0	0	0	0	0	0	0	-1	0	-1	$\pi_5 = \frac{v_s}{L\omega}$
π_6						1	0	0	0	0	0	0	0	0	-1	0	-2	$\pi_6 = \frac{g}{L\omega^2}$
π_7							1	0	0	0	0	0	0	0	-1	0	-2	$\pi_7 = \frac{A}{L\omega^2}$
π_8								1	0	0	0	0	0	0	0	0	1	$\pi_8 = T_d\omega$
π_9									1	0	0	0	0	0	-1	0	0	$\pi_9 = \frac{s}{L}$
π_{10}										1	0	0	0	0	0	0	0	$\pi_{10} = \theta$
π_{11}											1	0	0	0	0	0	0	$\pi_{11} = \varepsilon$
π_{12}												1	0	0	-1	-1	0	$\pi_{12} = \frac{\sigma}{L\gamma}$
π_{13}													1	0	-1	0	-1	$\pi_{13} = \frac{V}{L\omega}$
π_{14}														1	-1	0	-2	$\pi_{14} = \frac{a}{L\omega^2}$

We can get similarity criteria from Table 3.2, and further we can get similarity rule (respectively, see Tables 3.3 and 3.4).

Then we bring the association relation into Eq. (3.7) and we can get Eq. (3.9):

$$\begin{aligned}\pi_i = & \left[\frac{c}{L\gamma}\right]^{a_2} \cdot [\varphi]^{a_3} \cdot \left[\frac{E}{L\gamma}\right]^{a_5} \cdot [\mu]^{a_6} \cdot \left[\frac{V_s}{L\omega}\right]^{a_7} \cdot \left[\frac{g}{L\omega^2}\right]^{a_8} \cdot \left[\frac{A}{L\omega^2}\right]^{a_9} \cdot \\ & [T_d\omega]^{a_{10}} \cdot \left[\frac{S}{L}\right]^{a_{12}} \cdot [\theta]^{a_{13}} \cdot [\varepsilon]^{a_{14}} \cdot \left[\frac{\sigma}{L\gamma}\right]^{a_{15}} \cdot \left[\frac{V}{L\omega}\right]^{a_{16}} \cdot \left[\frac{a}{L\omega^2}\right]^{a_{17}}\end{aligned} \tag{3.9}$$

Comprehensively analyzing Table 3.4, we can know that some similar constants can be directly derived as follows:

$$C_\mu = 1 \quad C_\varphi = 1 \quad C_\varepsilon = 1 \quad C_\theta = 1$$

Table 3.3 Similarity criteria

$\pi_8 = T_d\omega$ $\pi_1 = c/(L\gamma)$ $\pi_9 = s/L$ $\pi_2 = \varphi$ $\pi_{10} = \theta$ $\pi_3 = E/(L\gamma)$ $\pi_{11} = \varepsilon$
$\pi_4 = \mu$ $\pi_{12} = \sigma/(L\gamma)$ $\pi_5 = V_s/(L\omega)$ $\pi_{13} = v/(L\omega)$ $\pi_6 = g/(L\omega^2)$ $\pi_{14} = a/(L\omega^2)$

Table 3.4 Similarity rule

$C_c/(C_l C_\gamma) = 1$ $C_\varphi = 1$ $C_E/(C_l C_\gamma) = 1$ $C_\mu = 1$ $C_{V_s}/(C_l C_\omega) = 1$ $C_g/(C_l C_\omega^2) = 1$
$C_A/(C_l C_\omega^2) = 1$ $C_\omega/C_{T_d} = 1$ $C_s = C_l$ $C_\theta = 1$ $C_\varepsilon = 1$ $C_\sigma/(C_l C_\omega) = 1$ $C_v/(C_l C_\omega) = 1$
$C_a/(C_l C_\omega^2) = 1$

Because geometric constant must equal the displacement constant, we can get $C_s = C_L$.

Gravitational acceleration g remains unchanged in the experiment, so we take $C_g = 1$, and we can know from the similarity rule in Table 3.4:

$$C_g/(C_l C_\omega^2) = 1, \tag{3.10}$$

$$C_A/(C_l C_\omega^2) = 1, \tag{3.11}$$

$$C_a/(C_l C_\omega^2) = 1 \tag{3.12}$$

We can get the similar constants $C_g = C_A = C_a = 1$

According to the similarity rule, we can get:

$$C_\omega = C_l^{-0.5} \tag{3.13}$$

Bringing Eq. (3.13) into the similarity rule, we can get:

$$C_\omega C_{T_d} = 1, \tag{3.14}$$

$$C_{V_s}/(C_l C_\omega) = 1, \tag{3.15}$$

$$C_v/(C_l C_\omega) = 1 \tag{3.16}$$

We can get the similarity rule:

$$C_{V_s} = C_V = C_l C_\omega = C_l^{0.5} \quad C_{T_d} = C_\omega^{-1} = C_l^{0.5}$$

The unit weight stays unchanged in the experiment, and we can take and get the similar constants:

$$C_E = C_c = C_l C_\gamma = C_l C_\sigma = C_l C_\gamma = C_l$$

Summarize the deduced results (see Table 3.5), in which the most important factor is geometric size. According to the practical situation of prototype and model, we choose geometric similarity ratio of 10:1 in the design.

3.1.4 Similar Materials

The correct selection of similar materials for shaking table test plays very important role in the success of the experiment. When selecting the similar materials, we not only need to make all their physical parameters meet the requirements of similarity,

Table 3.5 Similarity constants

No.	Physical parameters	Similarity relation	Similar constants	Notes
1	Geometric dimensions L	C_L	10	Controlled quantity
2	Unit weight γ	C_γ	1	Controlled quantity
3	Duration T_d	$C_{T_d} = C_L^{0.5}$	3.16	Controlled quantity
4	Cohesive force c	$C_c = C_L$	10	
5	Internal friction angle φ	$C_\varphi = 0$	1	
6	Dynamic modulus E	$C_E = C_L$	10	
7	Dynamic Poisson's ratio μ	$C_\mu = 1$	1	
8	shear wave velocity V_s	$C_{V_s} = C_L^{0.5}$	3.16	
9	Gravitational acceleration g	$C_g = 1$	1	
10	Input acceleration A	$C_A = 1$	1	
11	Frequency ω	$C_\omega = C_L^{-0.5}$	0.316	
12	Linear displacement s	$C_s = C_L$	10	
13	Angular displacement θ	$C_\theta = 1$	1	
14	Strain ε	$C_\varepsilon = 1$	1	
15	Response speed V	$C_V = C_L^{-0.5}$	10	
16	Stress σ	$C_\sigma = C_L$	3.16	
17	Response acceleration a	$C_a = 1$	1	

but also need to make them meet certain selecting principles. Based on years of research results of the predecessors, we summarize several principles that the similar rock materials shall meet:

(1) The raw materials shall be safe and harmless to humans;
(2) Try to buy the ordinary, accessible and low-price raw materials;
(3) The rock materials shall have stable chemical, mechanic, and physical character, and try to make them not affected by external conditions such as time, humidity, and temperature;
(4) The fabrication process of the similar materials needs to be as easy and convenient as possible;
(5) The finished similar materials need to have high insulation;
(6) The materials need to be plastic and easy for grinding and shaping, and have short solidification time;
(7) Adjusting the ration of materials could largely change their mechanic characteristics.

There are mainly two kinds of stratum lithology of rock slope in Wenchuan earthquake region: One is hard rock with representative of thick massive limestone and the other is soft rock with the representative of mudstone. The shaking model test mainly selects these two representative materials: hard rock and soft rock. The raw similarity materials used to configure the shaking table need to be of different type and performance. The model simulation materials are commonly composed of aggregate, cementing materials, and admixtures. According to the features of shaking table test and several commonly used similarity materials in model test as well as the mechanic performance indexes of various proportion ratios, this experiment adopts barite powder, river sand, gypsum, clay and water as raw materials, and the detailed performance and function of them are shown in Table 3.6.

According to the similarity rule, the density of the model material should be the same with that of the prototype. The proportional ratio of the similarity materials is unknown before we do the configuration. Referring to the similarity material proportional ratio and the commonly used similarity raw materials summarized by the similarity theory and the statics model test, and similarity materials and similar model, we regulate the main parameters of the similarity materials through different raw materials and adjusting their ratio to finally work out the proportional ratio of the similarity materials needed by the shaking table test: The proportional ratio of materials constituting soft rock is river sand: plaster: water: clay = 1.0:0.27:0.12:0.27; the proportional ratio of materials constituting hard rock is river sand: plaster: water: clay: barite powder = 1.0:0.5:0.28:0.60:0.80. For the concrete physical and mechanical indexes of similarity materials see Table 3.7, and for the photographs of laboratory configuration test see Figs. 3.2 and 3.3.

3.1.5 Test Instruments and Monitoring Points Disposition Principle

This experiment focuses on the responses of rock slope to earthquakes, and we mainly collect the acceleration and displacement response values of up and down parts of rock slopes under vibration loads, especially the acceleration responses. The acceleration sensor and laser displacement sensor are used to record the acceleration and displacement values of different parts of slopes, and

Table 3.6 Name and function of raw materials

Type	Name	Function
Aggregate	River sand	As coarse aggregate to change the internal friction angle of materials
Cementing materials	Barite powder	It is of high unit weight and can change the unit weight of materials. As filling materials, it is cohesive
Admixtures	Clay	It can preserve moisture and decrease mechanical properties
Binder	Water and gypsum	Different ratio can simulate different parameters. It plays the role of cementation

Table 3.7 Physical and mechanical parameters of hard rock and soft rock used in the experiment

Lithology	Density (g/cm^3)	Moisture content (%)	Cohesive force (kPa)	Internal friction angle (°)
Soft rock	2.2	9.655	104	41
Hard rock	2.3	7.826	200	45

Fig. 3.2 Strain-controlled direct shear apparatus

Fig. 3.3 Cutting-ring method for unit weight of soil test

high-frequency digital photography system is used to record the slope displacement changes during the earthquake.

3.1.5.1 Measurement Information and Selection of Sensor Types

The measurement indexes and monitoring contents are as follows: the acceleration and displacement values of different parts of slopes, and the slope displacement changes during the earthquake. After the experiment, analyze the slope response principle to earthquakes through comparison of acceleration, displacement, and deformation characteristics of different parts of the slopes.

(1) The acceleration sensor

The acceleration sensor mainly adopts the three-direction acceleration sensors produced by Donghua Test Technology Company co. Ltd, Jiangsu Province. Its measuring range is 0–20 m/s/s, weight 8 g, resonant frequency 12 kHz, sensitivity 60 mV/m/s/s (*X*, *Y*, *Z*), frequency range 0–1500 Hz (*X*), 0–1500 Hz (*Y*), 0–800 Hz (*Z*), transverse sensitivity ratio less than or equal to 5 % and external size 12 mm × 14 mm × 8 mm. In addition, in order to ensure the monitoring quality of acceleration data, this experiment adopts single-direction acceleration sensor on some key positions simultaneously: Its charge sensitivity is ~247 mV/ms^{-2}, resonant frequency is ~2.5 kHz, frequency range is 0.03 Hz–0.8 kHz, maximum transverse sensitivity is less than 5 %, the operating temperature is −20–80 °C, weight is about 230 g, and maximum measuring range is 20 m/s/s.

(2) Laser displacement sensor

The laser displacement sensor used the dynamic displacement sensor for earthquakes produced by Xi'an Bohua Measurement and Control electronic technology co., Ltd. Its detailed parameters are as follows: form: LVDT, measuring range: ±50 mm, frequency response: 0–500 Hz, accuracy: ≥0.2 %, output voltage: ±5 V, output interface: BNC interface, and motion parts: non-spring-rebound-type M4 thread. In addition, two laser displacement sensors were disposed on the outside of the model containers. Their measuring range is ±50 mm, sensitivity is 215.43 mV/mm, and operating temperature is −50–60 °C.

(3) High-frequency digital photography system

The high-frequency digital photography system was independently researched and developed by our research group, which could be used to monitor the displacement field of model in real time. At present, it has already formed complete hardware and software corollary equipment.

3.1.5.2 Monitoring Points Disposition Principle of Shaking Table Test

In the shaking table test, the disposition of the sensors should meet the following two principles:

(1) The sensors should be disposed in full reference to the dynamic response value calculation results of rock slopes under earthquake and the existing research results, and in close relation to the experiment purpose.
(2) With basic measurement information collection requirements of the experiment purpose met, the number of sensor should be as fewer as possible, so that to keep the completeness of models and avoid to make artificial geological structures due to instrument disposition so as not to affect the accuracy of research results.

3.1.6 Testing Points Disposition

In order to study the dynamic characteristics and dynamic response rules of rock slopes under strong earthquakes, different positions of rock slope models should arrange single-direction acceleration sensors, three-direction acceleration sensors, differential displacement sensors, and laser displacement sensors so as to get synchronous acceleration and displacement response data in the loading process. The single-direction acceleration sensors are charged with AC, and they are mainly used to test the acceleration field distribution of bedrock. Three-direction acceleration sensors are charged with DC, and they are mainly used to monitor the acceleration field inside the rock. Differential displacement sensors and laser displacement sensors are charged with AC and they are mainly used to monitor displacement changes in the earthquake. At the same time, in order to accurately describe the displacement changes of the whole model surface in the earthquake, this experiment adopts the independently researched and developed high-frequency digital photography system to monitor.

(1) Distribution of single-direction acceleration sensors

The single-direction acceleration sensors are mainly used to monitor the free-field acceleration response in the earthquake, and two of free-field points are arranged in this experiment. Three groups of acceleration monitoring point are laid out in free-field: one from top down and in free field two; only one group of acceleration monitoring point is laid out on the surface. See Figs. 3.4, 3.5, and 3.6 for detailed information.

(2) Distribution of three-direction acceleration sensors

Three-direction acceleration sensors are mainly used to monitor the acceleration amplification effect along elevation in the earthquake. In this experiment, four acceleration monitoring points are laid out in the central line on the 30°, 45°, 50°

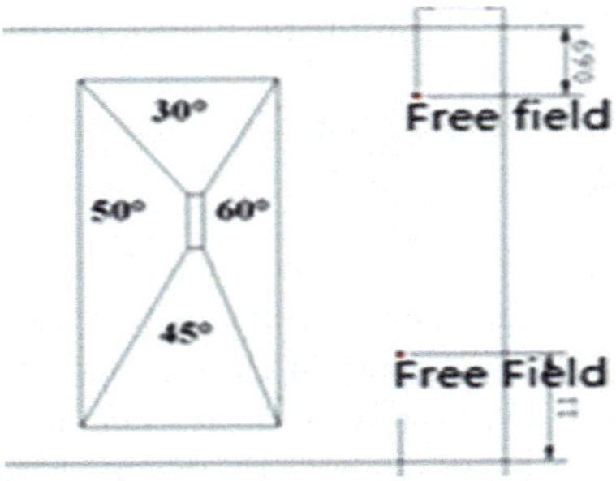

Fig. 3.4 Floor plan

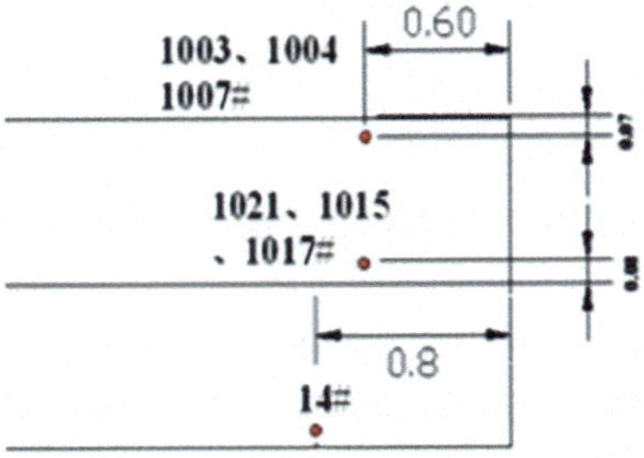

Fig. 3.5 Sectional drawing of free-field one

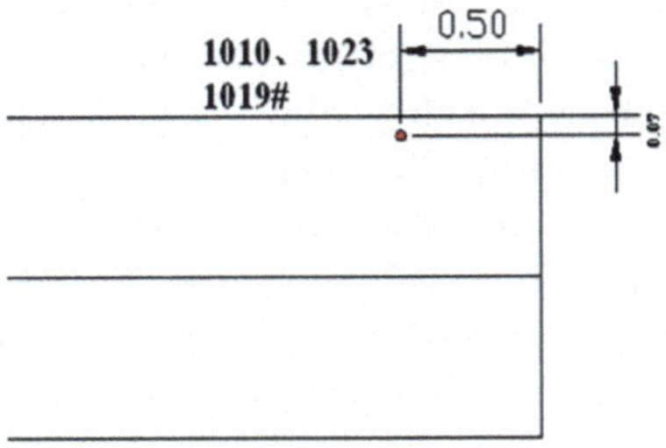

Fig. 3.6 Sectional drawing of free-field two

and 60° surface from top down. See Figs. 3.7, 3.8, 3.7, and 3.10 for detailed information.

(3) The arrangements of laser displacement sensors and differential displacement sensors

Laser displacement sensors and differential displacement sensors are mainly used to monitor the slope deformation in the earthquakes. In this experiment, one acceleration monitoring point was, respectively, laid out at the central line on the 30° and 60° surface, and a displacement sensor was laid out at the central line on the 50° surface.

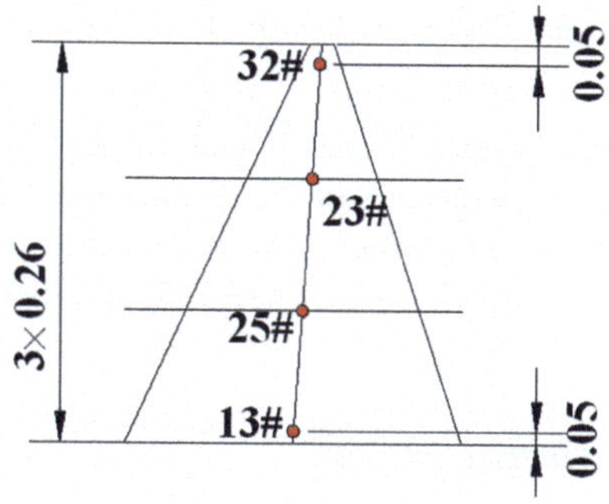

Fig. 3.7 Acceleration monitoring point arrangement plan on 30° surface

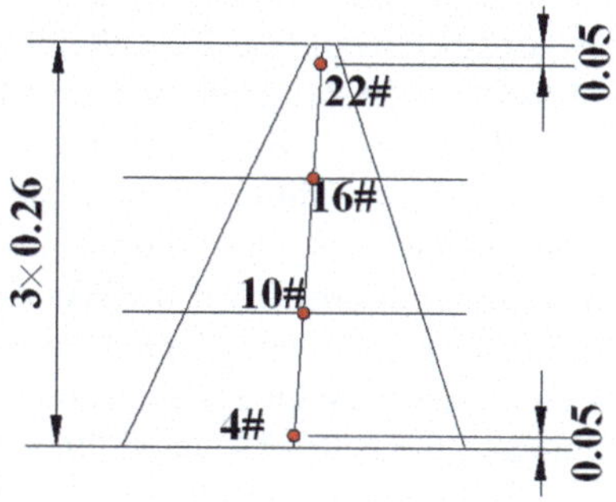

Fig. 3.8 Acceleration monitoring point arrangement plan on 45° surface

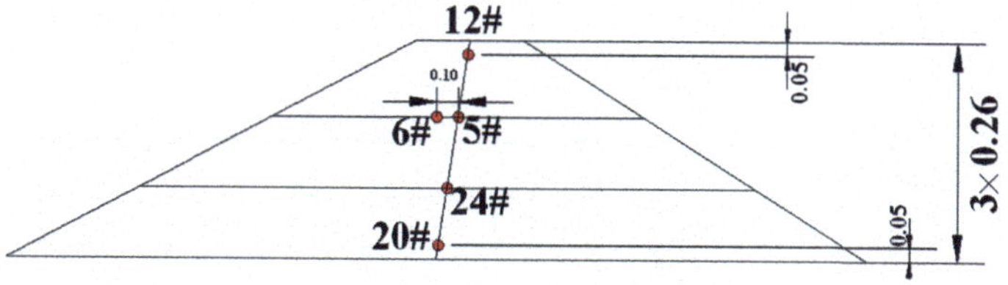

Fig. 3.9 Acceleration monitoring point arrangement plan on 50° surface

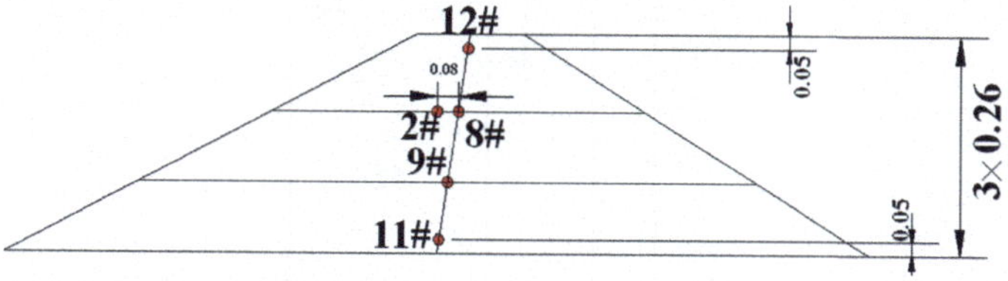

Fig. 3.10 Acceleration monitoring point arrangement plan on 60° surface

(4) The monitoring points arrangements of high-frequency digital photography system

High-frequency digital photography system was mainly used for research of the slope deformation in the earthquakes and the displacement field distribution of the whole model surface. In this experiment, the monitoring points are set up on the whole model surface with 0.2 m × 0.2 m network distribution.

3.1.7 Design and Manufacture

The factors influencing the slope dynamic response characteristics are various and complex, which include slope structure, topography, and geomorphology, lithologic character as well as loading conditions. However, a model with various influencing factors mixing together would be difficult for not only modeling but also analysis. Therefore, based on the principle to strengthen major factors, this experiment simplified the geologic body. In accordance with the model test purpose, this experiment designed slopes of different angles as characteristics to study the slope angle influence on seismic dynamic response rules.

For the two major kinds of rocks discovered in the Wenchuan earthquake region, i.e., hard rocks and soft rocks, this experiment adopts hard rocks with representative of limestone and soft rocks with representative of mudstone. Soft rocks simulate the slopes while hard rocks simulate the bedrocks. In addition, this experiment adopts homogeneous structure to simplify influencing factors. See Figs. 3.11 and 3.12 for detailed characteristic information of the model. In line with the similarity relations, the length scale of prototype and model is 10:1, and the major parameters of the model are as follows:

Fig. 3.11 Top view of shaking table test models

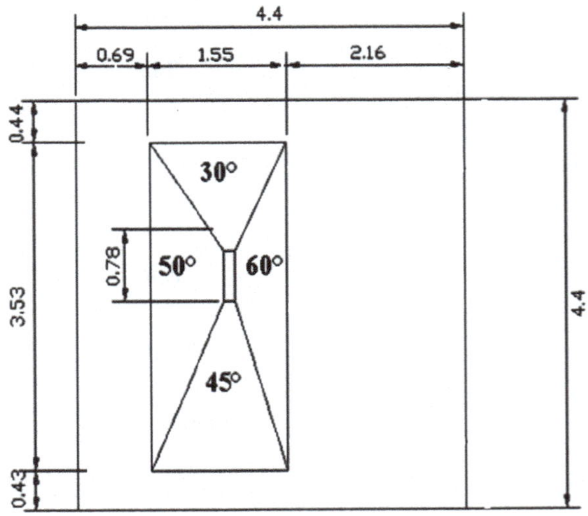

Fig. 3.12 Axonometric view of the finished shaking model test

① The total height of the model is 1.8 m, of which the height of the slope is 0.8 m, that of overlying soft rocks 0.5 m, and that of hard rocks 0.5 m. The length and width of the model are, respectively, 4.4 m;
② The length and width of the bottom line are, respectively, 3.53 and 1.55 m, and that of the top surface are, respectively, 0.78 and 0.1 m;
③ The model contains four slopes of different angles: 30°, 45°, 50°, and 60°;
④ The material composition of the model: The rock slopes are composed of soft rocks with physical parameters of mud-rocks as simulation; the bedrocks are composed of hard rocks with physical parameters of limestone as simulation.
⑤ The slope structures: The slopes inside the model container are composed of rocks of homogeneous materials.

3.1.7.1 Production and Transportation of Shaking Model Test Materials and Location of the Model Containers

This experiment used homogeneous materials and in situ casting to build test models. When building the models, we firstly prepared all the needed materials, of which river sand, barite powder, and plaster were all directly purchased from the manufactures, while clay was taken on site by our research group members. Then, we dealt with them with the process of smashing, grinding, and sifting. This experiment needs 20 tons of clay. See Figs. 3.13 and 3.14 for detailed dealing process. Besides, this experiment was carried out in the Seismic Laboratory of the second institute of Nuclear Power Institute of China, so the raw materials and model containers needed to be sent to the laboratory, and the model containers needed to be hoisted. See Figs. 3.15 and 3.16.

Fig. 3.13 Smashing of the clay

Fig. 3.14 Bagging of the smashed clay

Fig. 3.15 Loading and unloading of clay

Fig. 3.16 Location of the model containers

3.1.7.2 Model Filling of Shaking Table Test

As the materials were of large amount, this experiment used mixer for blending for 10 min, and when the materials mixed well, it used gantry crane filling them into the model container. See Fig. 3.17. The filling was by the measure of from top to bottom and layer by layer. After filled 20 cm each time, we conducted a density test and compacted it with vibrating by both machine and manpower, i.e., using manual work to vibrate at the boundary of the model and machine at the central parts so as to ensure the compaction quality. See Figs. 3.18 and 3.19. Besides, when building the models, people should embed the sensors in accordance with their designed location and try to avoid damage them. In order to prevent some sensors breaking down or being damaged in the vibrating, the embedding number should be over the designed number.

Fig. 3.17 Machine blending of the model materials

Fig. 3.18 Cutting-ring test for density

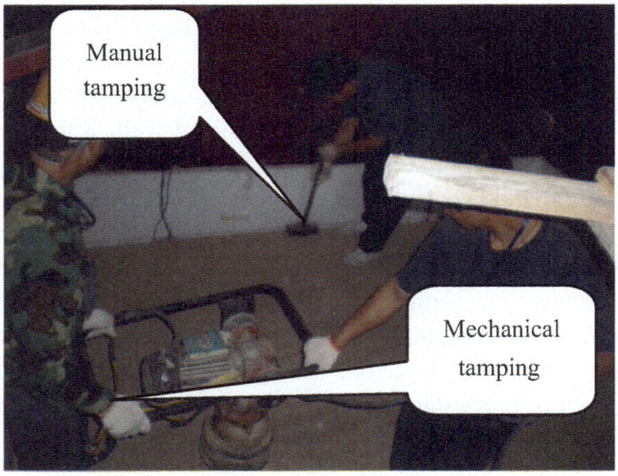

Fig. 3.19 Vibrating of the model materials

Fig. 3.20 Boundary of the test model

To avoid vibration wave reflecting at the model container border, it should eliminate the most of "boundary effect" influence on the experiment. Wave-absorbing materials of 30 mm thickness were added to the end walls in the vibrating direction, and foam was used as wave-absorbing materials (see Fig. 3.20).

3.1.8 Sensors

The correct installation and embedding of the test instrument plays a key role in the accuracy of the test data. Before embedding of the instruments, we need to use apparatus such as multimeter to detect and rectify them to ensure normal operation (see Fig. 3.21). In addition, as the model was built by blending and filling, water must be included in the similarity materials. In order to keep the normal operation of the sensors, they needed waterproofing work first. A layer of acid silicone glass sealant was coated on the sensors (see Fig. 3.22); before embedding the single-direction acceleration sensors, they needed to be put inside the insulated boxes, and the weight of the acceleration sensors and the boxes must be kept the same with that of the same volume of soil. The acceleration sensors and the boxes were connected with screws (see Fig. 3.23). Due to high demands on security direction by the three-direction acceleration sensors, so in order to prevent larger rotation of acceleration in the vibrating process, this experiment bonded a light sheet steel of high stiffness with size of 5 cm × 5 cm × 1 mm on the surface of the acceleration sensors. The embedding of the acceleration sensors adopted the measure of modeling and embedding at the same time (see Fig. 3.24). Displacement sensors were installed directly after the model was built. Before the installation, the brackets of the sensors needed to be fixed, and the brackets were made of i beam and were connected to the model containers. Holes were dug in the

Fig. 3.21 Rectification of three-direction acceleration sensors

Fig. 3.22 Seal of three-direction acceleration sensors

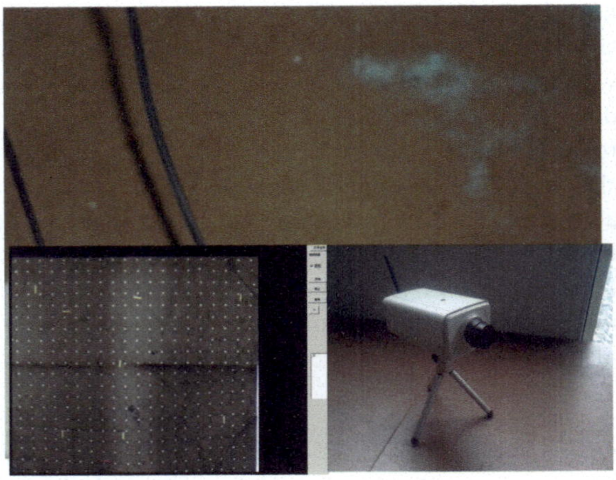

Fig. 3.23 Single-direction acceleration sensors

middle of the brackets, and the displacement sensors were fixed on the brackets through the holes with nuts. The end of the displacement sensor bar needed to be buried inside the slopes to ensure the needles would not slip (see Fig. 3.25). The laser displacement sensors were fixed three meters away from the shaking table, and baffle board was embedded on the slopes to reflect light beam of the laser displacement sensors (see Fig. 3.26). High-frequency digital photography system was fixed on the crane of the gantry crane at the top of the laboratory so as to reduce interference from the shaking table and increase test accuracy. At the same time, large amount of reflex attachments were set up evenly on the surface of the model surface, which are basically of the 0.2 m × 0.2 m network distribution (see Figs. 3.27 and 3.28).

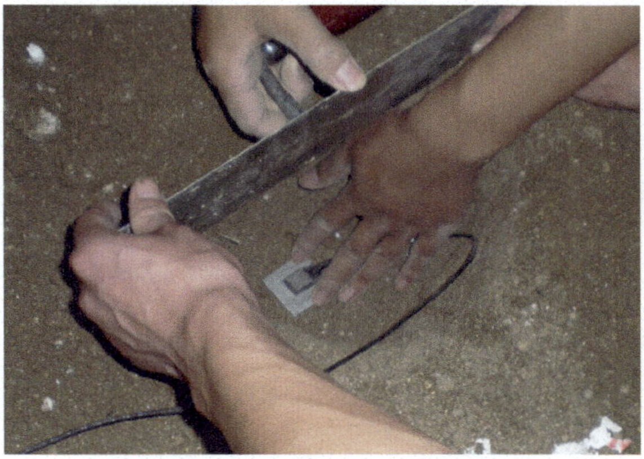

Fig. 3.24 Three-direction acceleration sensors

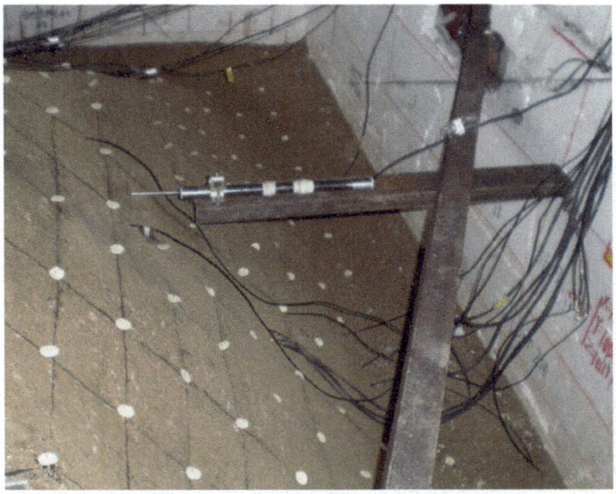

Fig. 3.25 Setup of differential displacement sensors

3.1.9 Loading Scheme

This experiment used input accelerogram to achieve earthquake wave simulation. According to the similarity rule, the seismic input accelerogram wave form of the model and the prototype are exactly the same; we only adjusted the time similarity ratio. This experiment aimed to study the influence of seismic wave type, frequency, and amplitude on the dynamic response rules of rock slopes, so we began by normalization processing of input seismic waves, and then exerted intensity by adjusting their amplitude to seven, eight, and nine degrees (modus earthquake,

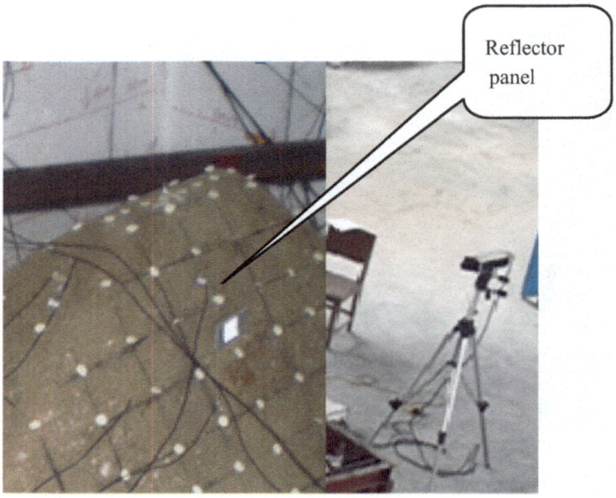

Fig. 3.26 Fix of laser displacement sensors

Fig. 3.27 High-frequency digital photography system

design earthquake, and rare earthquake), the three-direction earthquake time history. The major seismic types were Wenchuan wolong seismic waves, EL Centro seismic waves, and Kobe seismic waves. Figure 3.29 shows the three-direction seismic dynamic accelerogram of Wenchuan wolong seismic waves after the amplitude was adjusted to 0.1 g, and acceleration on the EW, NS, and UP directions was, respectively, input through the *X*, *Y*, and *Z* directions of the model (Fig. 3.12 specified this). Figure 3.30 shows the three-direction seismic dynamic accelerogram of EL Centro seismic waves, and acceleration on the EW, NS, and UP

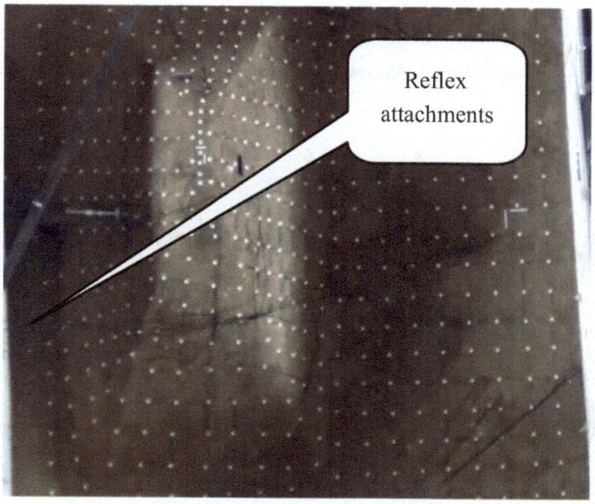

Fig. 3.28 Reflex attachments of high-frequency digital photography system

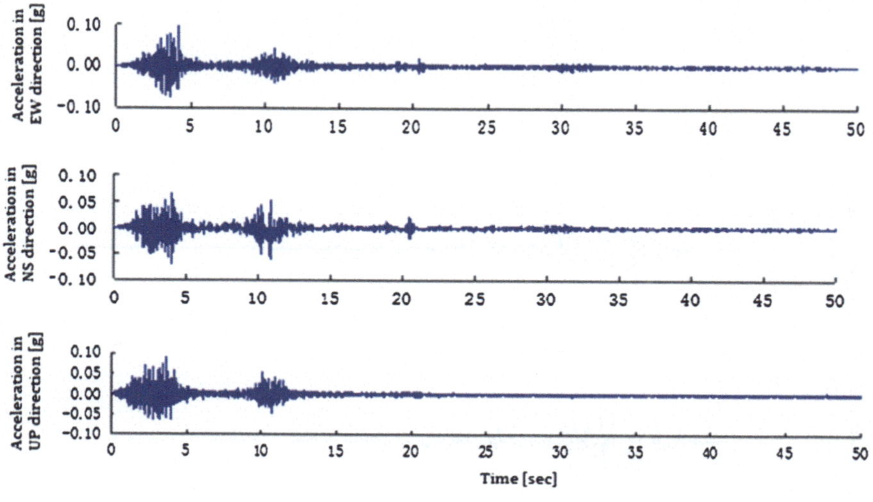

Fig. 3.29 Three-direction acceleration of Wenchuan wolong seismic waves after the amplitude was adjusted to 0.1 g

directions was, respectively, input through the *X*, *Y*, and *Z* directions. Figure 3.31 shows the three-direction seismic dynamic accelerogram of Kobe seismic waves, and acceleration on the EW, NS, and UP directions was, respectively, input through the *X*, *Y* and *Z* directions. In addition, in order to study the dynamic response rules under different seismic dynamic acceleration amplitudes and obtain the critical acceleration value when the slopes showed plastic deformation, the experiment

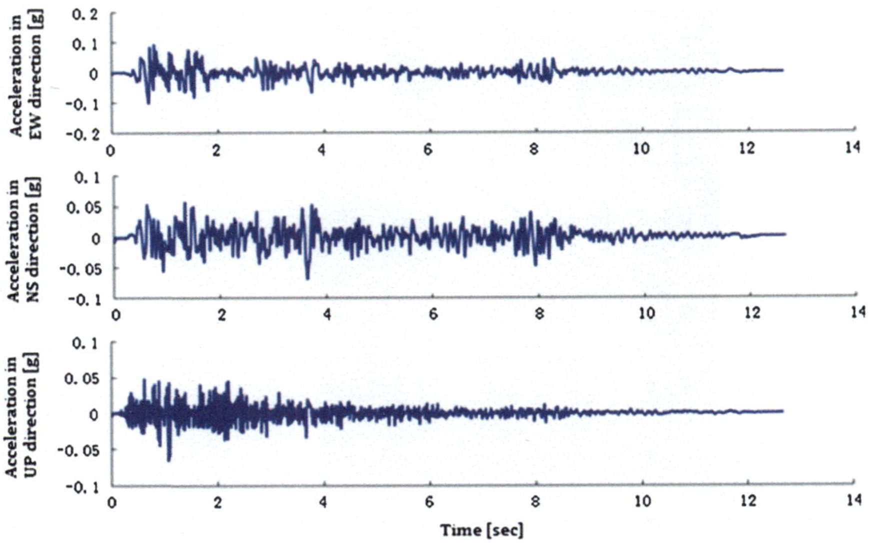

Fig. 3.30 Three-direction seismic dynamic accelerogram of EL Centro seismic waves after the amplitude was adjusted to 0.1 g

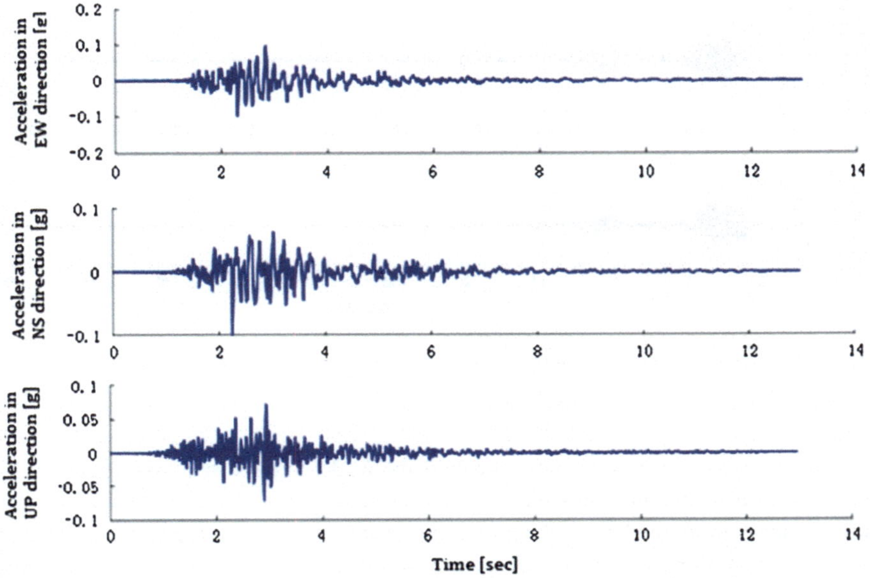

Fig. 3.31 Three-direction seismic dynamic accelerogram of Kobe seismic waves after the amplitude was adjusted to 0.1 g

planned to exert acceleration load of different amplitudes from small to large on the model.

White noise scanning is the key measure to test the dynamic characteristics of model structure. In order to know the dynamic characteristic changes of the model before and after vibration, this experiment used white noise scanning in different stages before and after vibration. It is initially decided that the first step was to load Kobe, EL Centro, and Wenchuan wolong seismic waves under seven-degree modus earthquake and then added amplitude gradually. See Table 3.8 for detailed loading scheme.

3.1.10 Data Collecting System

This experiment adopts two data collecting systems simultaneously. The NPIC adopts BBM data collecting system with 128 channels, and its maximum fiducial error is less than or equal to 0.5 %. Sensor signal conditioner was connected to electric charge transfer to convert the voltage signal, and its maximum fiducial error is less than or equal to 1 %. Data collecting, signal monitoring, and online analyzing were conducted at the same time (see Fig. 3.32). Southwest Jiaotong University adopted Donghua dynamic data acquisition instrument DH5923 with 32 channels, and the maximum sampling frequency was 5 kHz. It used 1394 or PCI interface, DMA measure, and computer communication to control, collect, store, and analyze in real time (see Fig. 3.33). In addition, it adopts high-frequency digital photography system to monitor the displacement field changes of the whole model site in real time. When model making, bracket setup, and sensor embedding all finished, we connected the sensors with the data collecting system and collect electric signals such as acceleration and displacement measured by the sensors (see Figs. 3.34 and 3.35).

The measurement indexes and observing contents are acceleration, displacement, and stress response value of different parts of the slopes as well as the deformation characteristics of the slopes. After the experiment, analyze the seismic response principle of slopes through comparison of acceleration, displacement, and deformation characteristics of different parts of the slopes.

3.1.11 Result Record

The experiment result records were composed of three major parts: the initial states of the test instrument models, experiment data, and the deformation of the test models.

Table 3.8 Loading scheme of shaking table test

Loading order: firstly 0.05 g white noise, then load one by one					
Loading scheme	PGA		Loading scheme	PGA	
1	7° modus earthquake	Kobe seismic wave	17	9° modus earthquake	Kobe seismic wave
2		El Centro seismic wave	18		El Centro seismic wave
3		Wenchuan wolong seismic waves	19		Wenchuan wolong seismic waves
4	0.05 g	White noise	20	0.05 g	White noise
5	7° design earthquake	Kobe seismic wave	21	9° design earthquake	Kobe seismic wave
6		El Centro seismic wave	22		El Centro seismic wave
7		Wenchuan wolong seismic waves	23		Wenchuan wolong seismic waves
8	0.05 g	White noise	24	0.05 g	White noise
9	8° modus earthquake	Kobe seismic wave	25	7° rare earthquake	Kobe seismic wave
10		El Centro seismic wave	26		El Centro seismic wave
11		Wenchuan wolong seismic waves	27		Wenchuan wolong seismic waves
12	0.05 g	White noise	28	0.05 g	White noise
13	8° design earthquake	Kobe seismic wave	29	8° rare earthquake	Kobe seismic wave
14		El Centro seismic wave	30		El Centro seismic wave
15		Wenchuan wolong seismic waves	31		Wenchuan wolong seismic waves
16	0.05 g	White noise	32	0.05 g	White noise
33	White noise 0.05 g				

Fig. 3.32 BBM data collecting system

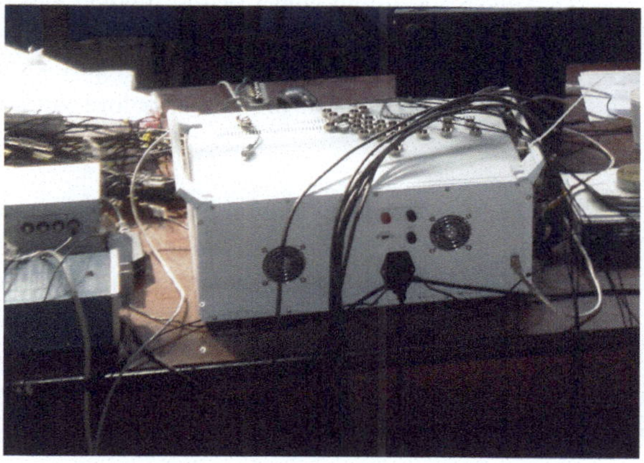

Fig. 3.33 DH5923 data collecting system

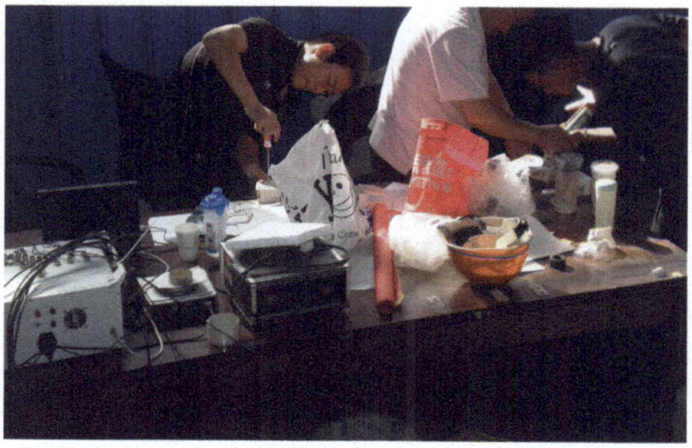

Fig. 3.34 Connection of sensors with data collecting systems

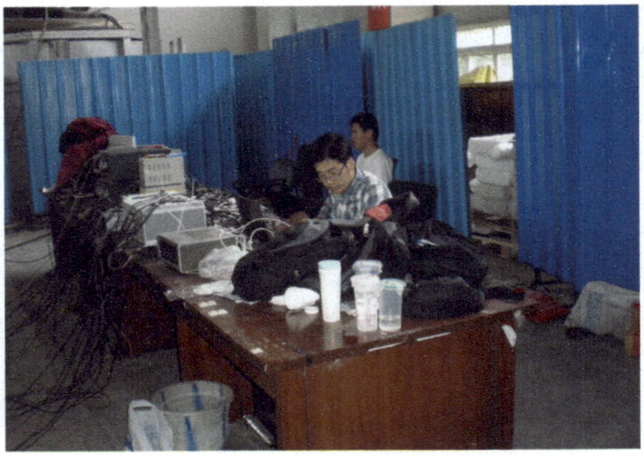

Fig. 3.35 Field debugging of shaking table test systems

(1) Before vibration test

① Check the initial states of the models and the preparation of all the test instruments, and debug the whole system.
② Use high-frequency digital photography system to record the initial states of different parts of the model before vibration test.

(2) During vibration test

① Set high-frequency digital photography system up above the shaking table test model to record the whole test process.
② after input of white noise and loading scheme of seismic wave of different levels, the computer data collecting system would automatically collect data about all the monitoring instruments during the vibration process.
③ Check the test model conditions and the working states of monitoring instruments.

(3) After vibration test

① Use steel ruler to measure the surface deformation and displacement of the model.
② Use camera to record the slope deformation and the crack developments.

3.2 Result Analysis

For test data of slope dynamic response rules experiment, acceleration is intuitive and easy to monitor and analyze, so many researchers see acceleration data as the main study object in the slope dynamic response rules experiment. This experiment arranged 19 three-direction acceleration test points inside the slopes; four of which are evenly set up from top down on the 30°, 45°, 50°, and 60° slope surface, and free-field acceleration test points were set up on both sides of the model. Then, through input of different seismic wave parameters, monitor the acceleration response values of different slope parts of different angles under earthquakes, analyze and monitor acceleration data, and study the influence of slope angles and ground motion parameters of seismic waves exerting on the slopes on the acceleration response rules.

For easy study on the acceleration response rules, the acceleration peak value was abbreviated as PGA, and acceleration amplification coefficient was defined as the ratio of acceleration peak value inside the slopes with that in the free field. Explain with acceleration on X axil for example: Suppose that the acceleration peak value on the X axil of point C inside the slope is AXC, and the acceleration peak value on the X axil of point D in the free field is AXD (see Fig. 3.36); then, the acceleration amplification coefficient δX on the X axil of this point could be represented as δX = ACX/ADX.

3.2.1 *Influence Factors of Acceleration Amplification Effect*

Difference of slope surface angle and seismic density as well as seismic ground motion input wave parameters has significant influence on the rock slope seismic ground motion characteristics. Therefore, this experiment build test model of bevel angles of 30°, 45°, 50°, and 60°, and input Wenchuan wolong, EL Centro, and Kobe seismic waves, which belong to modus earthquake, design earthquake, and rare earthquake with respective peak value of seven, eight, and nine degrees. Further, we researched on the influence of the above three conditions on the rock slope acceleration peak value elevation amplification effect.

Fig. 3.36 Diagram of the model

3.2.1.1 Effects of Bevel Angles on the Rock Slope PGA Elevation Amplification Effect

At present, a large number of research results show that bevel angles have significant influence on the seismic ground motion acceleration peak value elevation amplification effect. To clearly explain the influence of bevel angles on the seismic ground motion acceleration peak value elevation amplification effect, this chapter only selects peak values of test points on X, *Y*, and *Z* directions on slope surface of bevel angles of 30°, 45°, 50°, and 60° under the effect of Wenchuan wolong seismic waves with acceleration peak values of 0.1, 0.2, and 0.4 g. It is worth noticing that the model used in this experiment contains four surfaces, and if we analyze the traditional method, i.e., using global coordinate system with X, Y, and Z all directions, it will lead to that the free face of 30° and 45° surfaces faces the direction of 50° and 60° surfaces, so it will be difficult to study the influence of bevel angles on the seismic ground motion acceleration peak value elevation amplification effect.

Thus, this chapter selects two kinds of coordinate systems to study the acceleration peak value elevation amplification effect. One is the widely used global coordinate system with *X*, *Y*, and *Z* all directions, which is used to simply explain the above problems, and the other is local coordinate system, whose three directions are the free face direction of the slopes (*L*), the strike of the slopes (*M*), and the vertical direction (*N*). The detailed information of these two coordinate systems is shown in Fig. 3.37. See Figs. 3.38 and 3.39 for analysis results under global coordinate system and Figs. 3.40 and 3.41 for that under local coordinate system. It is noteworthy that the free face direction (*L*) in the local coordinate system is not the outer normal direction of the slopes, but the horizontal direction pointing to the outside of the slopes. The vertical direction *N* is the same with the *Z* direction in the global coordinate system and the slope strike is the extension direction of slope in the space.

Fig. 3.37 Diagram of coordinate system changes

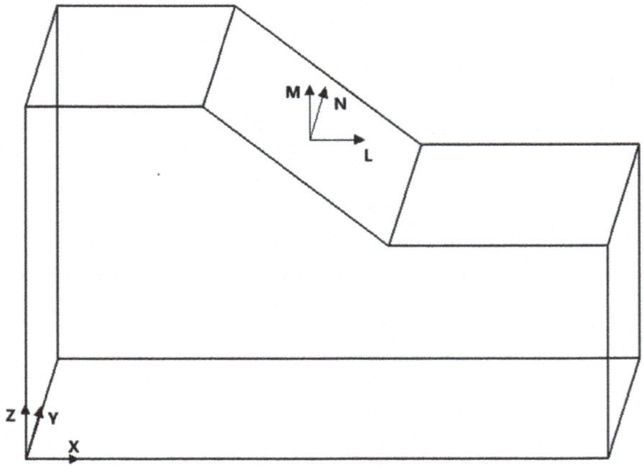

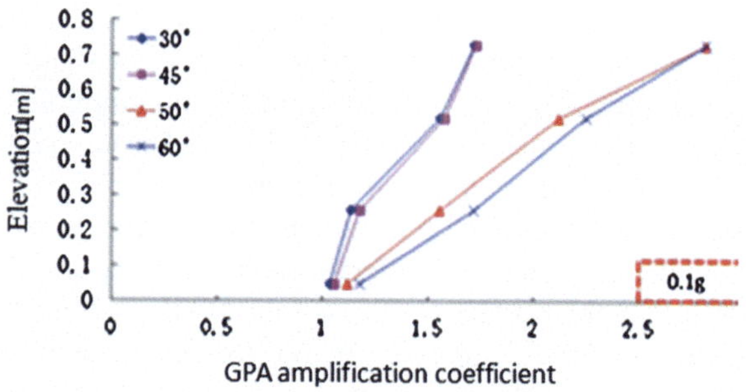

Fig. 3.38 PGA elevation amplification coefficient in *X* direction

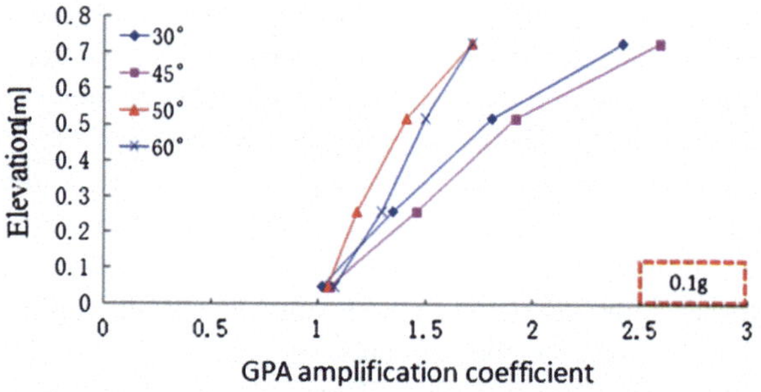

Fig. 3.39 PGA elevation amplification coefficient in *Y* direction

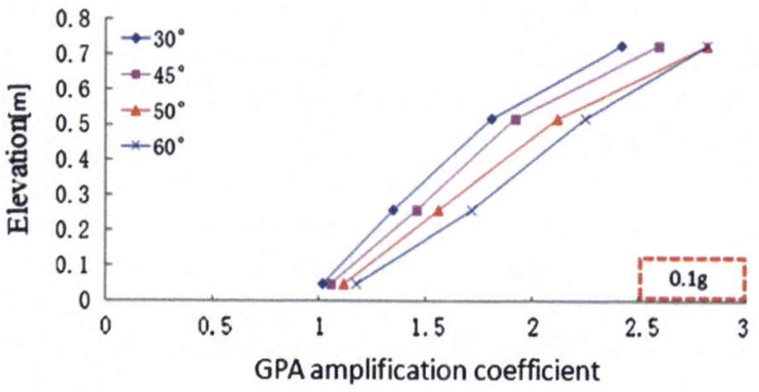

Fig. 3.40 PGA elevation amplification coefficient in *L* direction

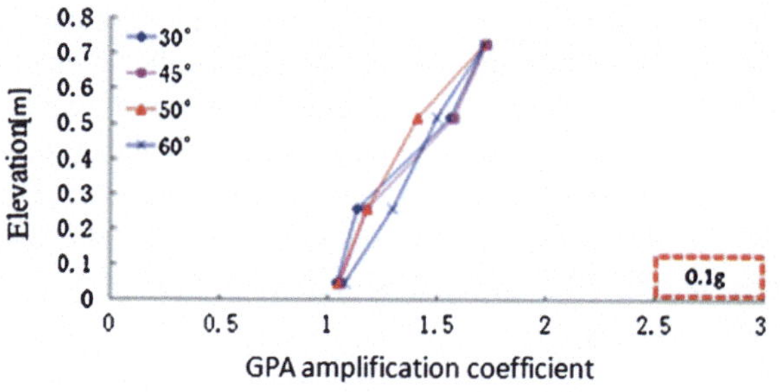

Fig. 3.41 PGA elevation amplification coefficient in *M* direction

We can see from Figs. 3.38 and 3.39 that under the effect of Wenchuan wolong seismic waves with PGA = 0.1 g, the PGA of *X* and *Y* directions on the surfaces of bevel angles of 30°, 45°, 50°, and 60° all amplified to different extent along elevation. For acceleration on the *X* axil, the acceleration peak value amplification rules of different slope surfaces are as follows: 60° surface > 50° surface > 45° surface > 30° surface. For acceleration on the Y axil, the acceleration peak value amplification rules of different slope surfaces are as follows: 45° surface > 30° surface > 60° surface > 50° surface. While as Figs. 3.40 and 3.41 show that under local coordinate system(*L*, *M*, *N*), for acceleration on the *L* axil, the acceleration peak value amplification rules of different slope surfaces are as follows: 60° surface > 50° surface > 45° surface > 30° surface. For acceleration on the M axil, the acceleration peak value amplification rules of different slope surfaces are as follows: 60° surface, 50° surface, 45° surface and 30° surface are basically the same. It can thus be seen that the results under global coordinate system and the local coordinate system are inconsistent. This may be because the *X*(*Y*) direction is the free face direction for 50° and 60° surfaces(*Y* direction is the slope strike), and the slope strike for 30° and 45° surfaces (*Y* is the free face direction). However, it is unilateral to compare test results of slope free face surface with that of the slope strike to research the acceleration peak value elevation amplification effect as it is difficult to clearly study on the influence of bevel angles in the acceleration peak value elevation amplification effect. Therefore, this dissertation used local coordinate system to elaborate the acceleration peak value elevation amplification effect.

We can see from Figs. 3.40, 3.41, 3.42, 3.43, 3.44, 3.45, 3.46, 3.47, and 3.48 that, under the effect of Wenchuan wolong seismic waves with PGA = 0.1, 0.2, and 0.4 g, the PGA of free face direction (*L*), slope strike (*M*), and vertical direction (*N*) on the surfaces of bevel angles of 30°, 45°, 50° and 60° all amplified to different extent along elevation. For acceleration on *L* direction, the acceleration amplification rules of all surfaces are as follows: 60° surface > 50° surface > 45° surface > 30° surface. For acceleration on *M* direction, the acceleration amplification rules of all surfaces are as follows: The amplification effects of 60° surface, 50° surface, 45° surface and 30° surface are basically the same. For acceleration on *N* direction, the acceleration amplification rules of all surfaces are as follows: 60° surface > 50° surface > 45° surface > 30° surface. Therefore, with the increase of the bevel angles, the PGA elevation amplification effect in free face direction and vertical direction will strengthen gradually, while that in slope strike would basically stay the same. For easy explanation for the above phenomenon, this chapter will simplify the problem to plane strain problem and explain from the angle of seismic wave transmission characteristics. It can be known from elastic wave theory that the vertically incidenting SV wave will come across wave field separation when it reached the slope surface, and it will separate SV-reflected wave of the same type and P-reflected wave of new type (see Fig. 3.49). Potential function of SV incident wave, SV-reflected wave, and P-reflected wave is as follows: potential function of SV incident wave: $\varphi_i = E_s e^{i(wt - k_1 x + k_{2s} x_3)}$; potential function of *P*-reflected wave:

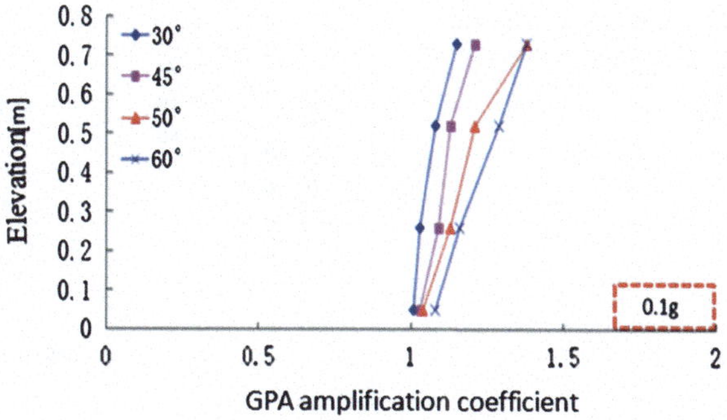

Fig. 3.42 PGA elevation amplification coefficient in vertical direction

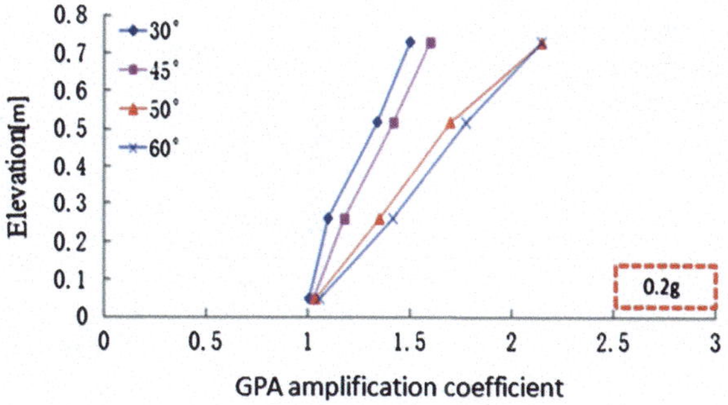

Fig. 3.43 PGA elevation amplification coefficient in free face direction

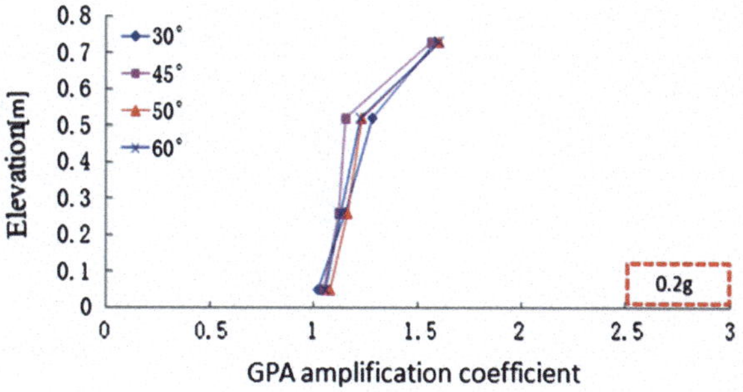

Fig. 3.44 PGA elevation amplification coefficient in slope strike

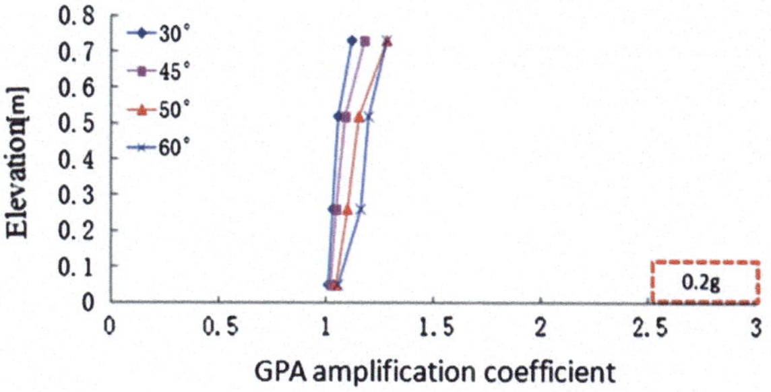

Fig. 3.45 PGA elevation amplification coefficient in vertical direction

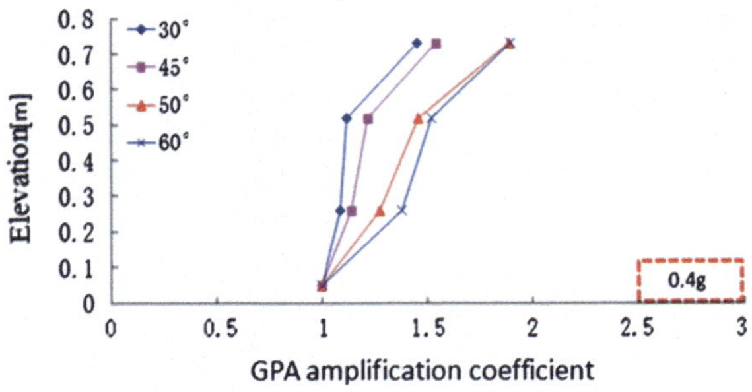

Fig. 3.46 PGA elevation amplification coefficient in free face direction

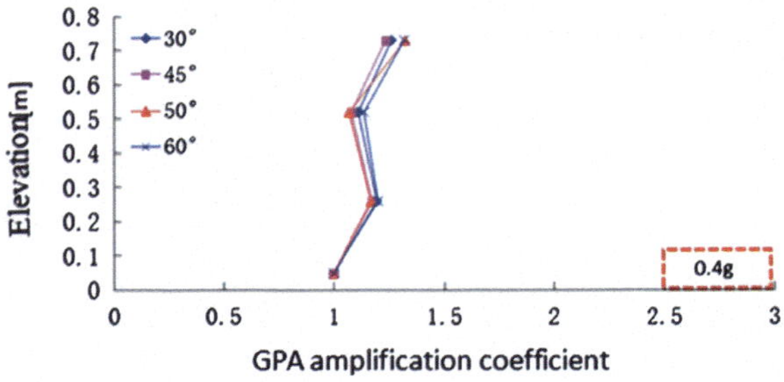

Fig. 3.47 PGA elevation amplification coefficient in slope strike

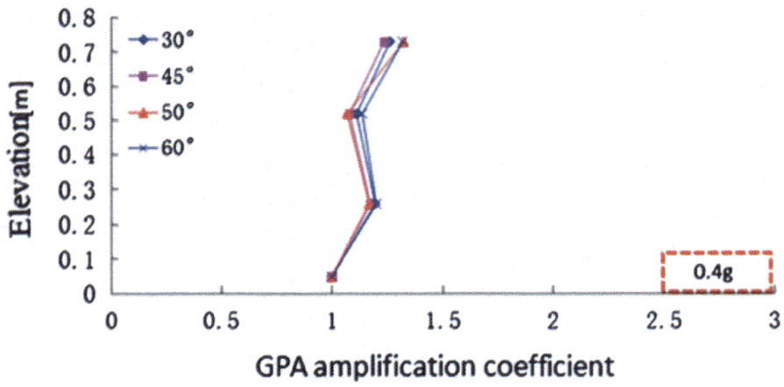

Fig. 3.48 PGA elevation amplification coefficient in vertical direction

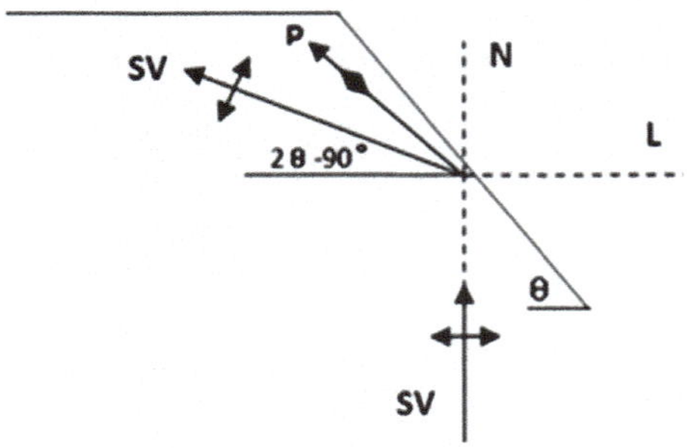

Fig. 3.49 Wave field separating phenomenon of SV incident wave

$$\varphi_r = F_p e^{i(wt - k_1 x_1 - k_{2p} x_3)};$$

potential function of SV-reflected wave:

$$\varphi = E_s e^{i(wt - k_1 x_1 + k_{2S} x_3)} + F_s e^{i(wt - k_1 x_1 - k_{2S} x_3)};$$

φ is the incident angle.

$$k_1 = \frac{\omega \sin \varphi}{V_s},$$

$$k_{2s} = \begin{cases} \sqrt{\left(\frac{w}{V_s}\right)^2 - k_{1s}^2} & k_{1s} \le \frac{w}{V_s} \\ -i\sqrt{\left(\frac{w}{V_s}\right)^2 - k_{1s}^2} & k_{1s} > \frac{w}{V_s} \end{cases} \quad k_{2p} = \begin{cases} \sqrt{\left(\frac{w}{V_p}\right)^2 - k_{1p}^2} & k_{1p} \le \frac{w}{V_p} \\ -i\sqrt{\left(\frac{w}{V_p}\right)^2 - k_{1p}^2} & k_{1p} > \frac{w}{V_p} \end{cases}$$

$$\frac{F_P}{E_S} = \frac{-2V_p^2 \sin 2\varphi \cos 2\varphi}{V_s^2 \sin 2\theta \sin 2\varphi + V_p^2 \cos 2\varphi} \quad \frac{F_s}{E_S} = \frac{V_s^2 \sin 2\theta \sin 2\varphi - V_p^2 \cos 2\varphi}{V_s^2 \sin 2\theta \sin 2\varphi + V_p^2 \cos 2\varphi},$$

Based on the above function, bringing the relevant parameters of 30°, 45°, 50°, and 60° surface into it to solve, the F_P/E_S ratio becomes negative and the F_S/E_S ratio becomes positive, which means the phase position of SV incident wave is the same with SV-reflected wave, but opposite to *P*-reflected wave. The further explanation is that the vibration component of SV incident wave and SV-reflected wave is same in *L* and *N* directions, while vibration component of SV incident wave and *P*-reflected wave is opposite in *L* and *N* directions. In addition, the vibration component of SV incident wave is E_s and 0 in *L* and *N* directions, the vibration component of SV-reflected wave is $(E_S + F_S)\sin(2\theta - 90°)$ and $(E_S + F_S)\cos(2\theta - 90°)$ in *L* and *N* directions, and the vibration component of *P*-reflected wave in *L* and *N* direction is $F_p\cos(2\theta 1 - 90°)$ and $F_p\sin(2\theta 1 - 90°)$, so the total vibration in *L* and *N* directions is $E_s + (E_S + F_S)\sin(2\theta - 90°) + F_p\cos(2\theta 1 - 90°) - (E_S + F_S)\cos(2\theta - 90°) + F_p\sin(2\theta 1 - 90°)$. Besides, with the increase of incident angle θ, the reflection angle of SV wave increases gradually, and its vibration direction is gradually close to *L* direction, while the vibration component of *P* wave in *L* direction decreases with its reflection angle $\theta 1$ increases; i.e., $(E_S + F_S)\sin(2\theta - 90°) + F_p\cos(2\theta 1 - 90°) - (E_S + F_S)\cos(2\theta - 90°) + F_p\sin(2\theta 1 - 90°)$ increases gradually. Therefore, with the increase of the bevel angles, the PGA elevation amplification effect in free face direction and vertical direction will strengthen gradually. In line with the above thoughts, the PGA elevation amplification effect in slope strike will come to the same conclusion.

In order to reveal more clearly the influence of bevel angles on PGA in *L*, *M*, and *N* directions, this chapter selects test point of *H*/4 to the slope top as research object, and detailed research results are shown in Figs. 3.50, 3.51, and 3.52.

It can be known from Figs. 3.49, 3.50, and 3.51 that with the increase of bevel angles, the acceleration amplification coefficient in free face direction, slope strike, and vertical direction all increases gradually, and distributed in "step-form"; i.e., the amplification coefficient of 30°–45° and 50°–60° increased slightly, while that of 45°–50° increased suddenly. In addition, the change of PGA amplification coefficient with step heights is as follows: free face direction > vertical direction > slope

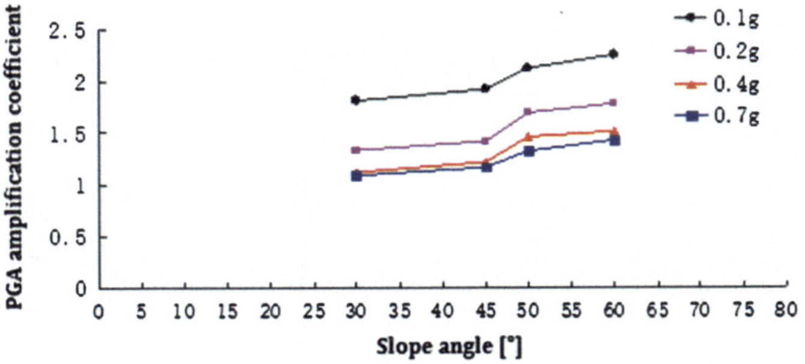

Fig. 3.50 PGA elevation amplification coefficient in free face direction

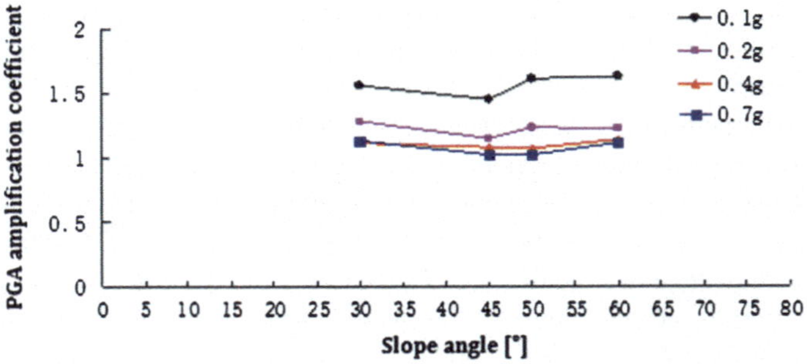

Fig. 3.51 PGA elevation amplification coefficient in slope strike

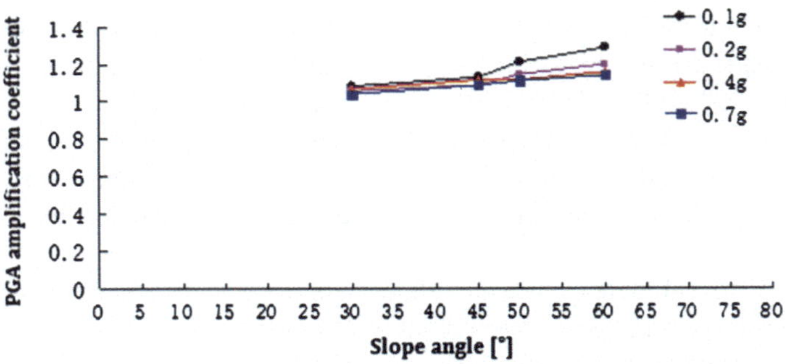

Fig. 3.52 PGA elevation amplification coefficient in vertical direction

strike, and with the effects of Wenchuan wolong seismic waves with peak value of 0.1, 0.2, 0.4 and 0.7 g, the PGA amplification coefficient of 50° surface is, respectively, 42.3, 19.7, 19.7, and 13.8 % higher that of 45° surface. The above analysis results illustrate that with the increase of bevel angles, the PGA amplification coefficient did not increase all the same, but with two turning points; i.e., 45° surface is the sudden increase turning point of PGA amplification coefficient and 50° surface is the leveling off turning point of PGA amplification coefficient. Therefore, the seismic dynamic response of rock slopes with bevel angles above 45° will be apparently higher of that below 45°, which is identical with the results of the Wenchuan earthquake damage survey of "The Landslide Disaster Occurs Mainly in Slopes with Bevel Angles above 40°." In addition, the above test results will be elaborated from the perspective of theoretical calculation in Chap. 5 "theoretical explanation of time and frequency analysis of two-sided rock slope elevation amplification effects."

From all the above analysis, it can be known that, with the increase of bevel angles, the PGA elevation amplification effects in free face direction and vertical direction will rise gradually, while at the same time, there exists a sudden increase in turning point at bevel angles of 45° and a levelling off turning point at 50°. However, the PGA elevation amplification effects in slope strikes stay unchanged with the increase of bevel angles, and its steps are comparatively gentle.

3.2.1.2 Effects of Seismic Density on the Rock Slope PGA Elevation Amplification Effect

To clearly explain the influence of seismic density on the seismic ground motion acceleration peak value elevation amplification effect, this chapter only selects peak values of test points on *L, M,* and *N* directions on slope surface of bevel angles of 30°, 45°, 50°, and 60° under the effect of Wenchuan wolong seismic waves with acceleration peak values of 0.1, 0.2, 0.4, and 0.7 g. See Figs. 3.53, 3.54, 3.55, 3.56, 3.57, 3.58, 3.59, 3.60, 3.61, 3.62, 3.63, and 3.64 for detailed calculation results.

We can see from Figs. 3.53, 3.54, 3.55, 3.56, 3.57, 3.58, 3.59, 3.60, 3.61, 3.62, 3.63, and 3.64 that, under the effect of Wenchuan wolong seismic waves with PGA = 0.1, 0.2, and 0.7 g, the PGA amplification effects of free face direction (*L*), slope strike (*M*), and vertical direction (*N*) on the surfaces of bevel angles of 30°, 45°, 50°, and 60° are identical to the changes of seismic wave peak values, i.e., PGA = 0.1 g > PGA = 0.2 g > PGA = 0.4 g > PGA = 0.7 g. Therefore, with the increase of PGA input, the PGA elevation amplification effect in free face direction, vertical direction, and slope strike will decrease gradually. In order to reveal more clearly the influence of bevel angles on PGA in *L, M,* and *N* directions, this chapter selects test point of *H*/4 to the slope top as research object, and detailed research results are shown in Figs. 3.65, 3.66, and 3.67.

It is known from Figs. 3.65, 3.66 and 3.67 that with the increase of input ground motion PGA, the acceleration elevation amplification effect decreases gradually and shows characteristics of magnitude saturation. Magnitude saturation characteristics

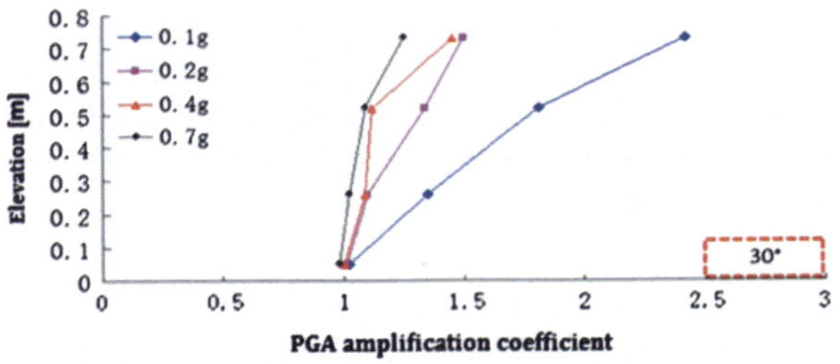

Fig. 3.53 PGA elevation amplification coefficient in free face direction

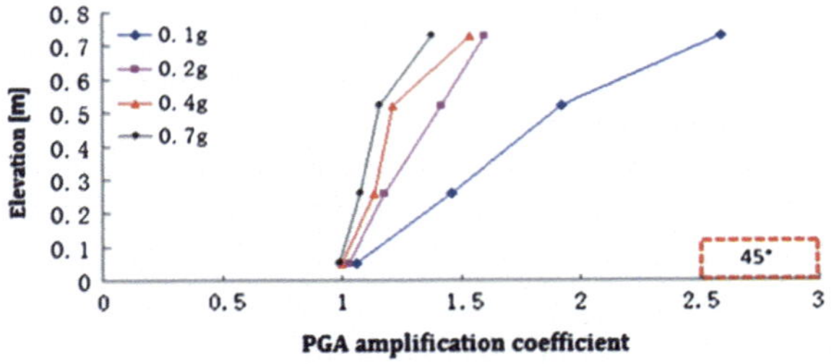

Fig. 3.54 PGA elevation amplification coefficient in free face direction

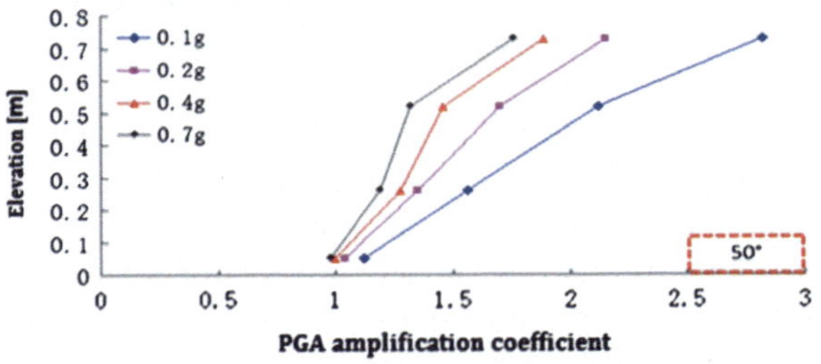

Fig. 3.55 PGA elevation amplification coefficient in free face direction

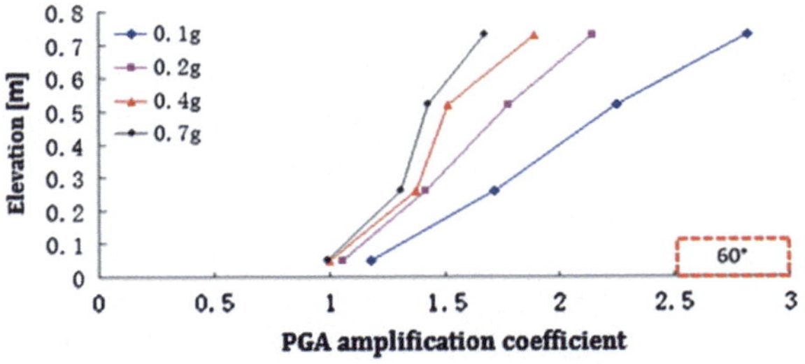

Fig. 3.56 PGA elevation amplification coefficient in free face direction

Fig. 3.57 PGA elevation amplification coefficient in slope strike

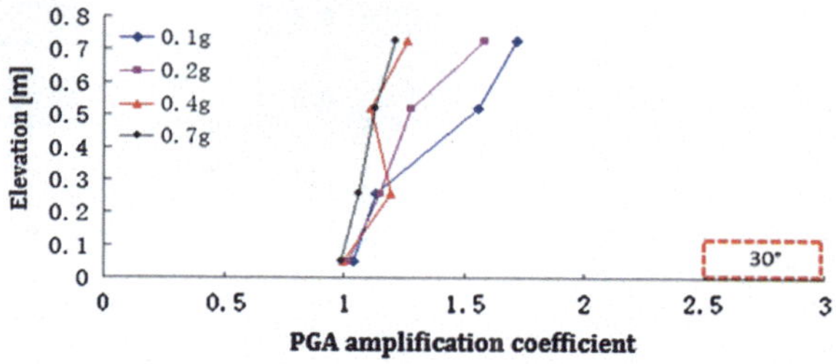

Fig. 3.58 PGA elevation amplification coefficient in slope strike

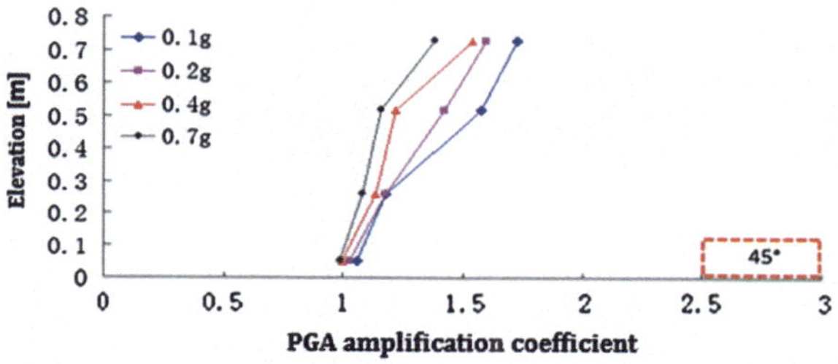

Fig. 3.59 PGA elevation amplification coefficient in slope strike

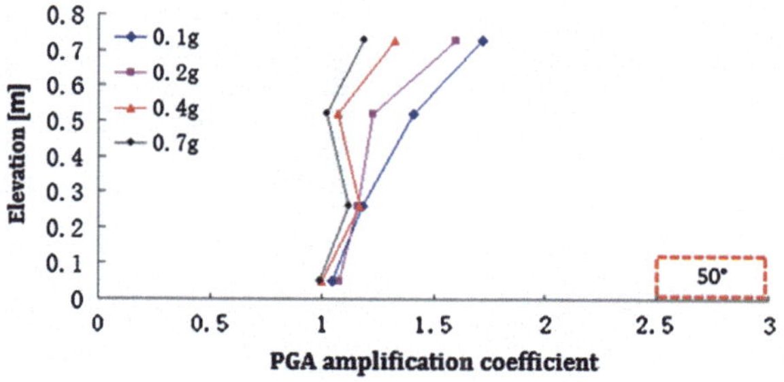

Fig. 3.60 PGA elevation amplification coefficient in slope strike

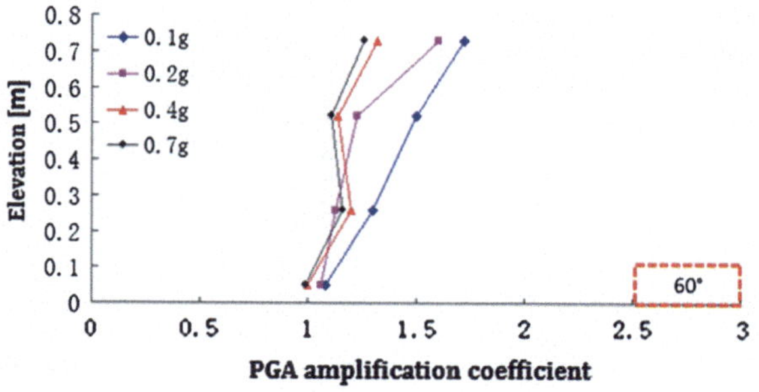

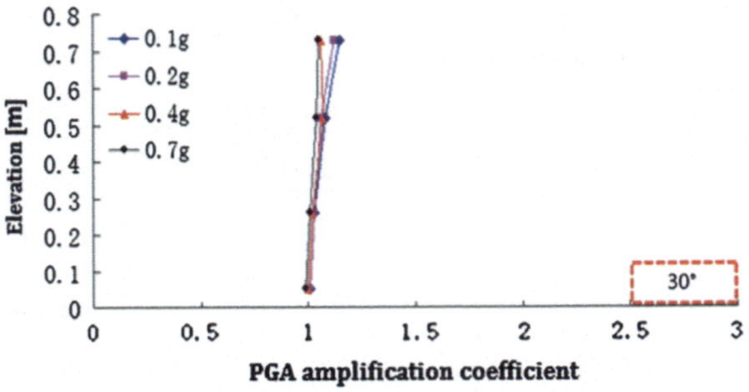

Fig. 3.61 PGA elevation amplification coefficient in vertical direction

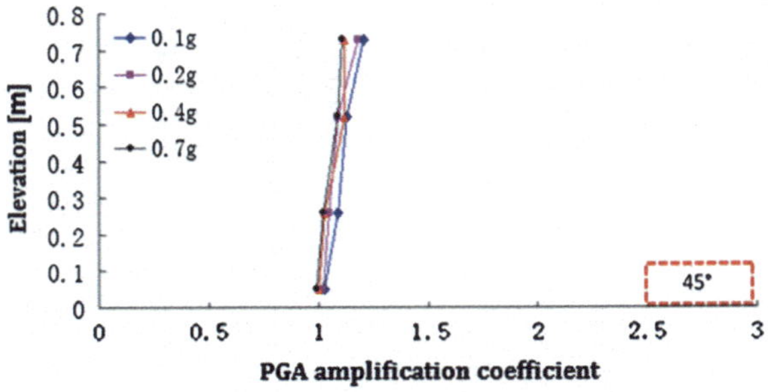

Fig. 3.62 PGA elevation amplification coefficient in vertical direction

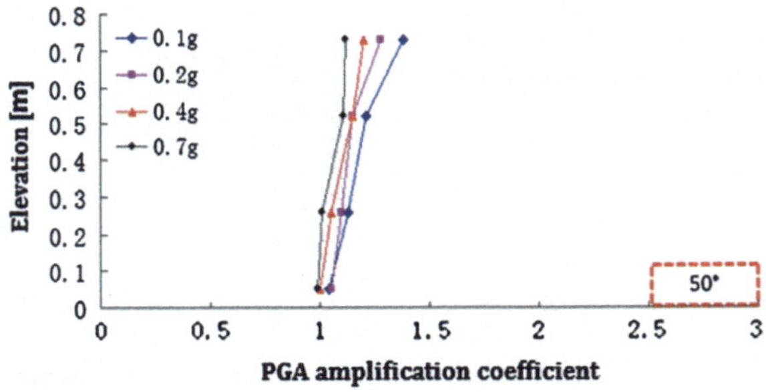

Fig. 3.63 PGA elevation amplification coefficient in vertical direction

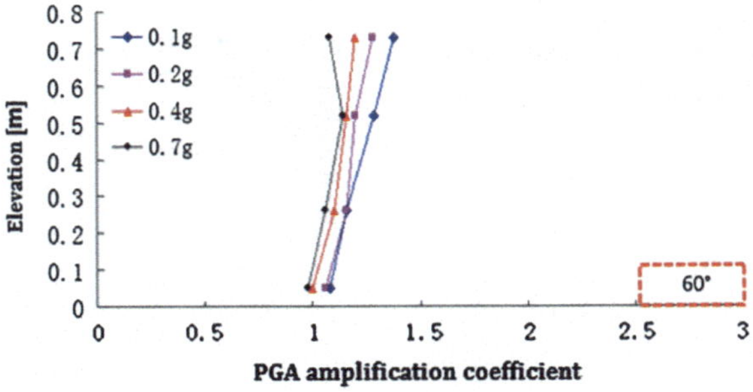

Fig. 3.64 PGA elevation amplification coefficient in vertical direction

mainly refer to that with the input ground motion PGA, the dynamic shear strength and modulus of soils declined, damping ratio of soils increases, nonlinear characteristics strengthen gradually, and energy consumption of seismic waves enlarges, which causes the attenuation of acceleration amplification effect, and the increase of rock and earth mass shock insulation and shock absorption effects.

From the above analysis, we can know that with increase of input ground motion PGA, the acceleration elevation amplification effect decreases gradually and shows characteristics of magnitude saturation.

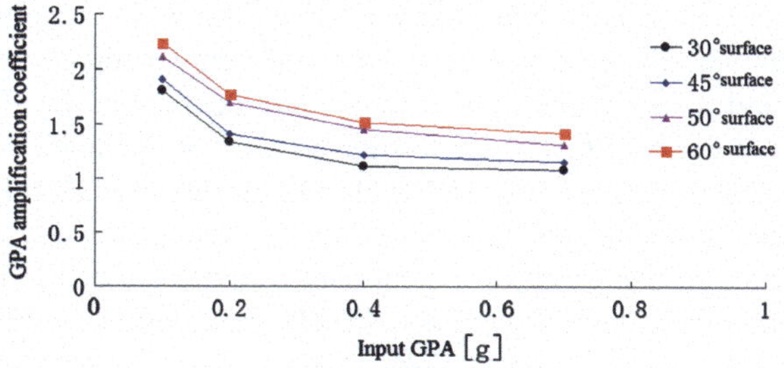

Fig. 3.65 PGA elevation amplification coefficient in free face direction

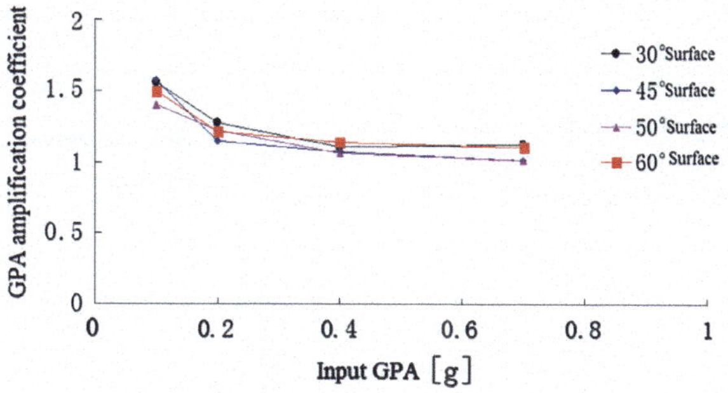

Fig. 3.66 PGA elevation amplification coefficient in slope strike

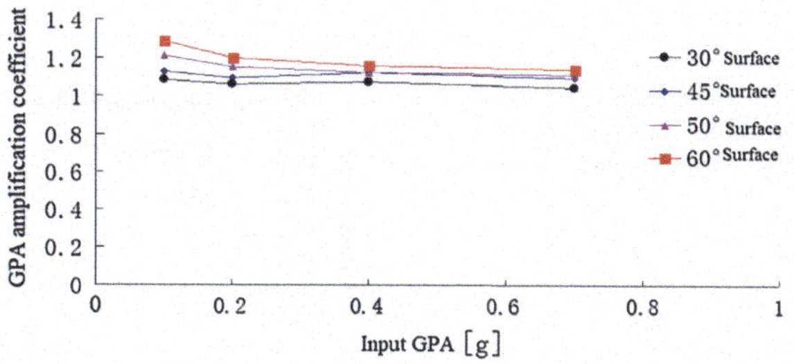

Fig. 3.67 PGA elevation amplification coefficient in vertical direction

3.2.1.3 Effects of Seismic Wave Types on the PGA Elevation Amplification Effect

To clearly explain the influence of seismic wave types on the seismic ground motion PGA elevation amplification effect, this chapter only selects peak values of test points on three directions on slope surface of bevel angles of 30°, 45°, 50°, and 60° under the effect of Wenchuan wolong, EL Centro, and Kobe seismic waves with acceleration peak values of 0.1 g. See Figs. 3.68, 3.69, 3.70, 3.71, 3.72, 3.73, 3.74, 3.75, 3.76, 3.77, 3.78, and 3.79 for detailed results.

We can see from Figs. 3.68, 3.69, 3.70, 3.71, 3.72, 3.73, 3.74, 3.75, 3.76, 3.77, 3.78 and 3.79 that under the effect of Wenchuan wolong, EL Centro, and Kobe seismic waves with PGA = 0.2 g, the PGA amplification effects of free face direction (*L*), slope strike (*M*), and vertical direction(N) on the surfaces of bevel angles of 30°, 45°, 50°, and 60° are identical to the changes of input seismic wave types, i.e., EL Centro seismic wave > Wenchuan wolong seismic wave > Kobe seismic wave. It also shows that PGA elevation amplification coefficient in vertical direction changes little related to the seismic wave-type changes. The above results may be due to that the dominant frequencies of EL Centro seismic wave in three directions are, respectively, 4.6 Hz (*X* direction), 4.49 Hz (*Y* direction), and 6.5 Hz (*Z* direction), those of Kobe seismic wave are, respectively, 6.84 Hz (*X* direction), 6.60 Hz (*Y* direction), and 8.09 Hz (*Z* direction), and those of Wenchuan wolong seismic wave are, respectively, 4.24 Hz (*X* direction), 3.56 Hz (*Y* direction), and 6.23 Hz (*Z* direction). Using mode analysis method, we can get that the first natural

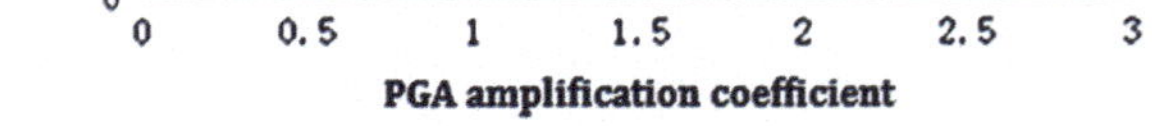

Fig. 3.68 PGA elevation amplification coefficient in free face direction

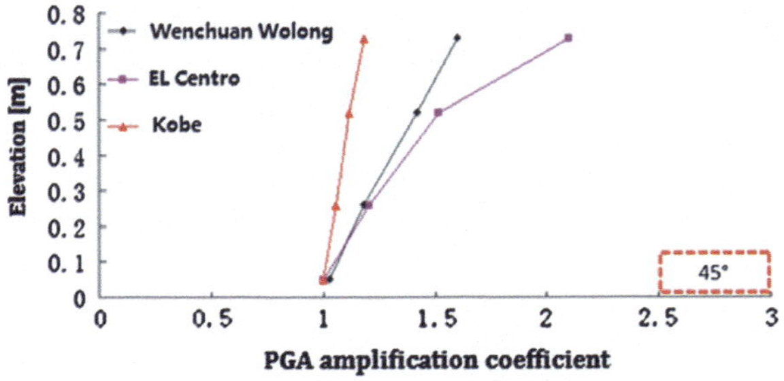

Fig. 3.69 PGA elevation amplification coefficient in free face direction

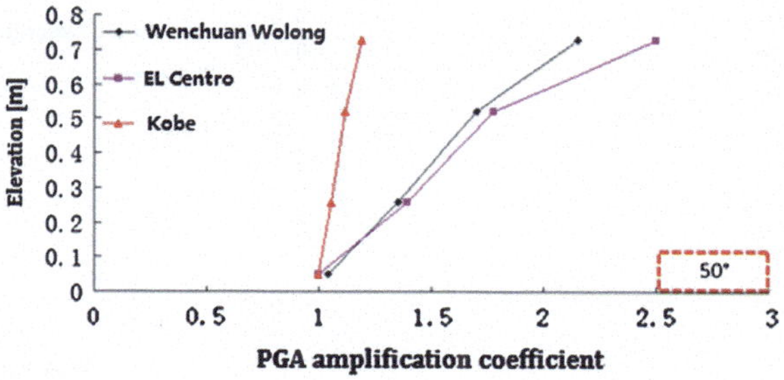

Fig. 3.70 PGA elevation amplification coefficient in free face direction

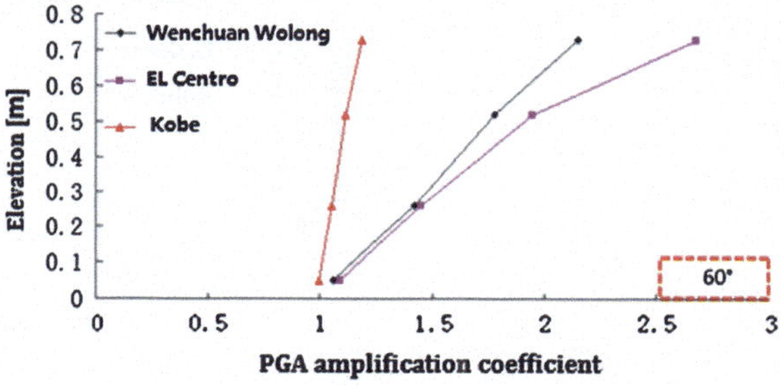

Fig. 3.71 PGA elevation amplification coefficient in free face direction

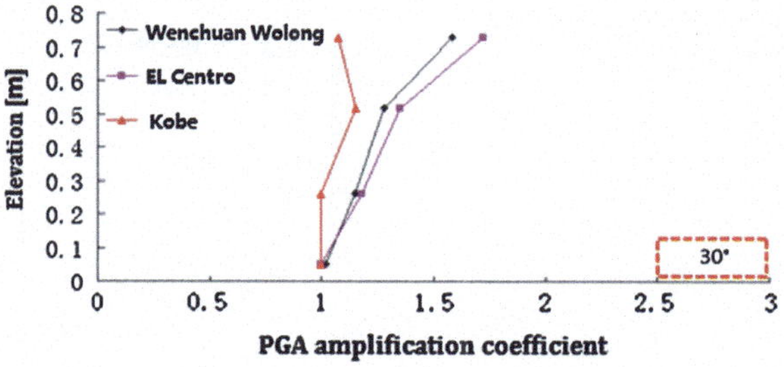

Fig. 3.72 PGA elevation amplification coefficient in slope strike

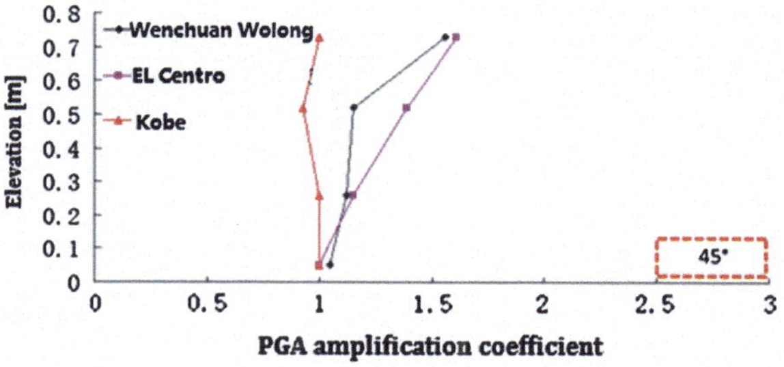

Fig. 3.73 PGA elevation amplification coefficient in slope strike

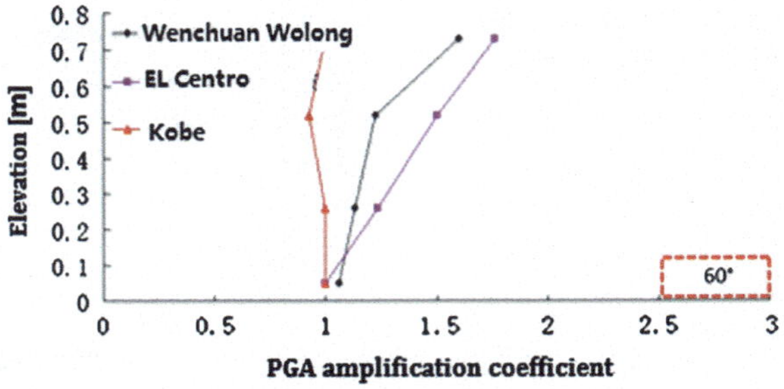

Fig. 3.74 PGA elevation amplification coefficient in slope strike

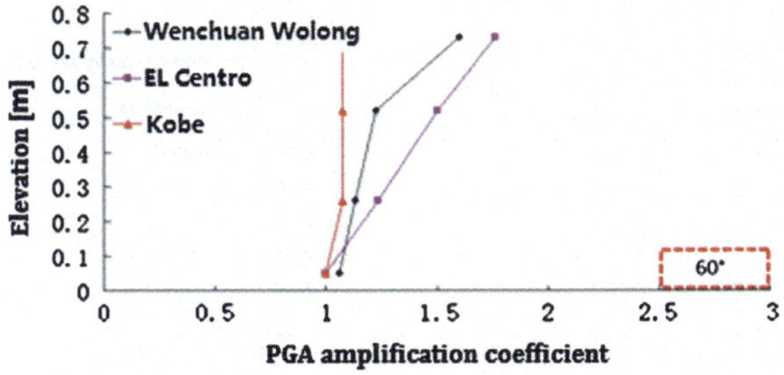

Fig. 3.75 PGA elevation amplification coefficient in slope strike

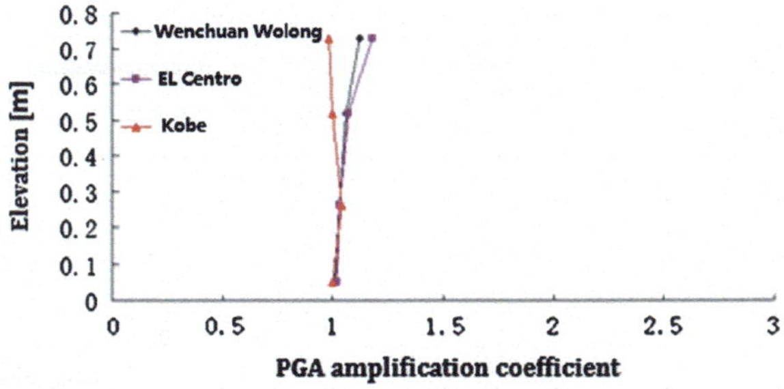

Fig. 3.76 PGA elevation amplification coefficient in vertical direction

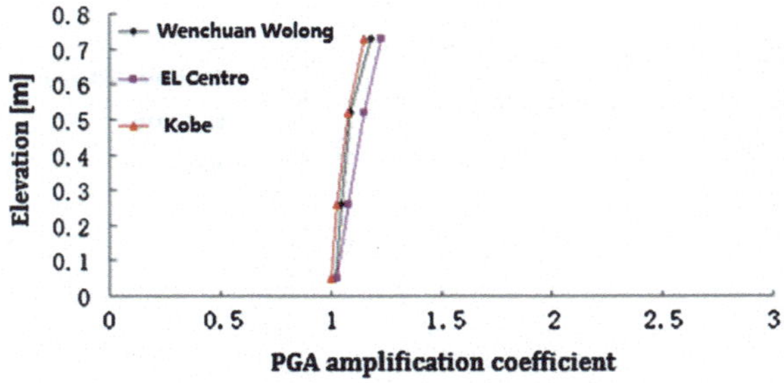

Fig. 3.77 PGA elevation amplification coefficient in vertical direction

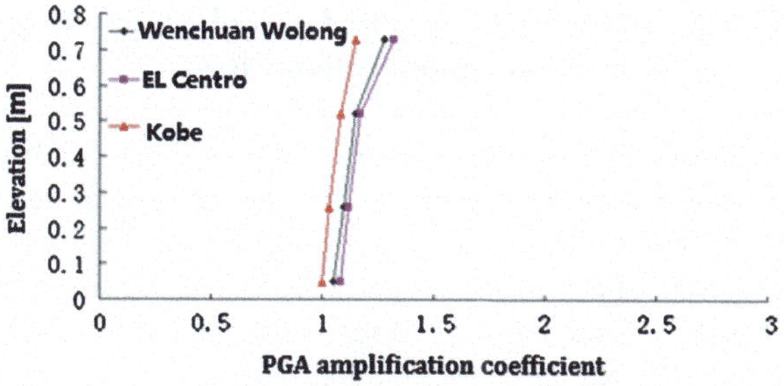

Fig. 3.78 PGA elevation amplification coefficient in vertical direction

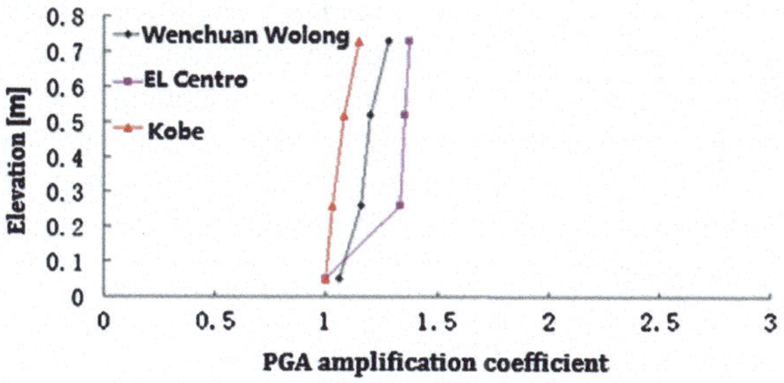

Fig. 3.79 PGA elevation amplification coefficient in vertical direction

frequency of the test model is 4.5 Hz, and the dominant frequency of EL Centro seismic wave on *X* and *Y* directions is the closest to it, so it is easy to cause resonance. The second close to it is Wenchuan wolong seismic wave and the last one is Kobe seismic wave. In addition, the dominant frequency of all three kinds of seismic waves in vertical direction all differentiates a lot to the first natural frequency of the test model. Therefore, the seismic wave types have little influence on the PGA amplification effect in vertical direction.

While it is noteworthy that the EL Centro, Wenchuan wolong, and Kobe seismic wave used in this chapter are all acceleration versus time data of same-elevation free-field test points at the bottom of the slopes, but not the input seismic waves at the bottom of shaking model test, which is mainly because the spectral components of seismic ground motion get changed when it is transmitted from the model container bottom through the soil layer of the slope bottom field upward, and what has real influence on slope dynamic response should be seismic waves directly input through the slope bottom, but not seismic ground motion input through the bottom of model container. Besides, this experiment did not set up monitoring points at the bottom of slope bottoms, but at the free field of same elevation. Therefore, this chapter selects acceleration versus time data of same-elevation free-field test points at the bottom of the slopes to analyze.

Based on all the above analysis, the seismic wave types have significant importance on rock slope acceleration elevation amplification effects, which is mainly exemplified in relation between the dominant frequency of seismic waves and the natural frequency of slopes.

3.2.1.4 Influence of Seismic Ground Motion Response Direction PGA Elevation Amplification Effect

To clearly explain the influence of seismic ground motion direction on the PGA elevation amplification effect, this chapter only selects peak values of test points on *L*, *M*, and *N* directions on slope surface of bevel angles of 30°, 45°, 50°, and 60° under the effect of Wenchuan wolong seismic waves with acceleration peak values of 0.1, 0.2, and 0.4 g. See Figs. 3.80, 3.81, 3.82, 3.83, 3.84, 3.85, 3.86, 3.87, 3.88, 3.89, 3.90, and 3.91 for detailed results.

We can see from Figs. 3.80, 3.81, 3.82, 3.83, 3.84, 3.85, 3.86, 3.87, 3.88, 3.89, 3.90, and 3.91 that under the effect of Wenchuan wolong seismic waves with PGA = 0.1, 0.2, and 0.4 g, the PGA amplification effects of free face direction (*L*), slope strike (*M*), and vertical direction (*N*) on the surfaces of bevel angles of 30°, 45°, 50°, and 60° are basically identical, i.e., free face direction > slope strike > vertical direction, which shows apparent "free face direction amplification effects." The reason for above results may be that the seismic waves came across reflection and refraction and formed complex wave field which caused larger acceleration.

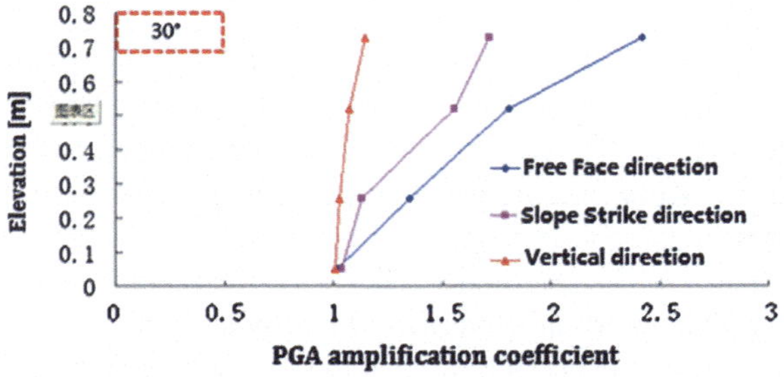

Fig. 3.80 PGA elevation amplification coefficient (PGA = 0.1 g)

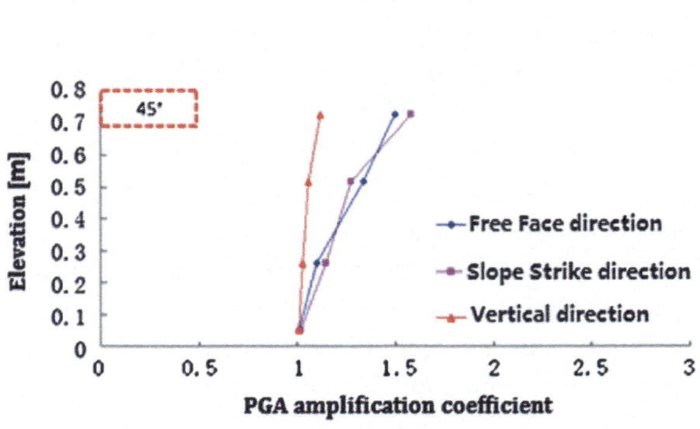

Fig. 3.81 PGA elevation amplification coefficient (PGA = 0.1 g)

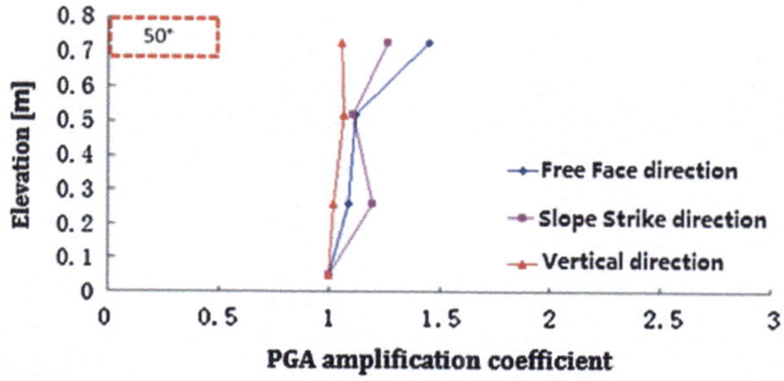

Fig. 3.82 PGA elevation amplification coefficient (PGA = 0.1 g)

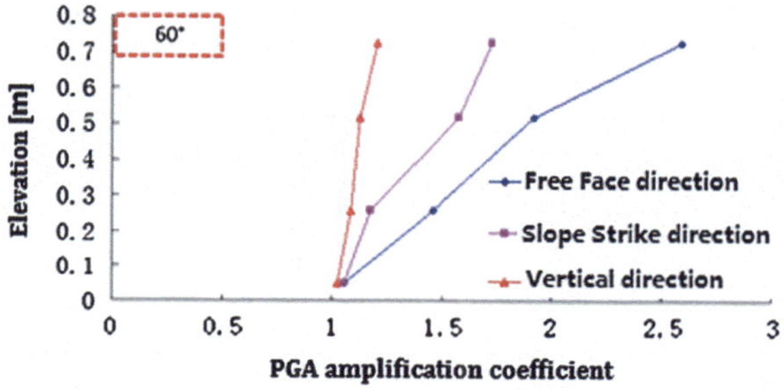

Fig. 3.83 PGA elevation amplification coefficient

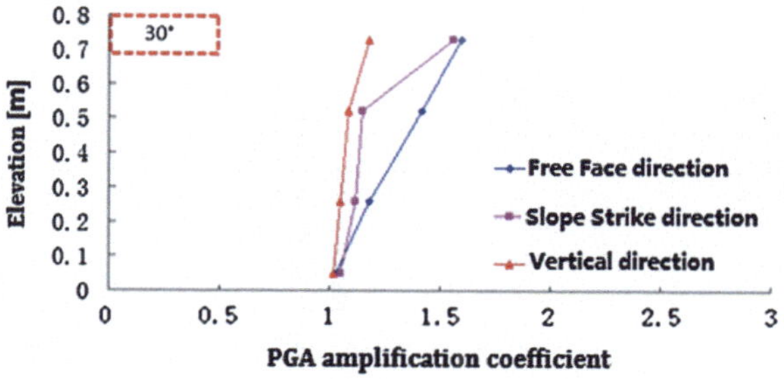

Fig. 3.84 PGA elevation amplification coefficient (PGA = 0.2 g)

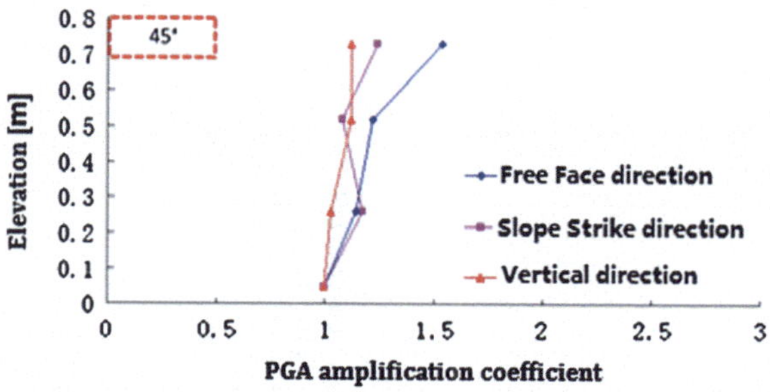

Fig. 3.85 PGA elevation amplification coefficient (PGA = 0.2 g)

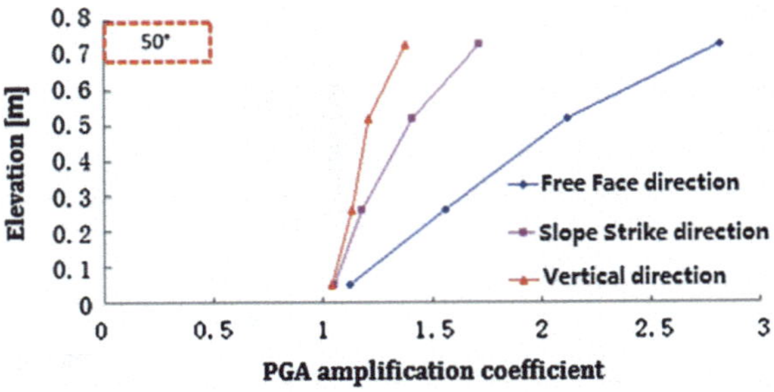

Fig. 3.86 PGA elevation amplification coefficient (PGA = 0.2 g)

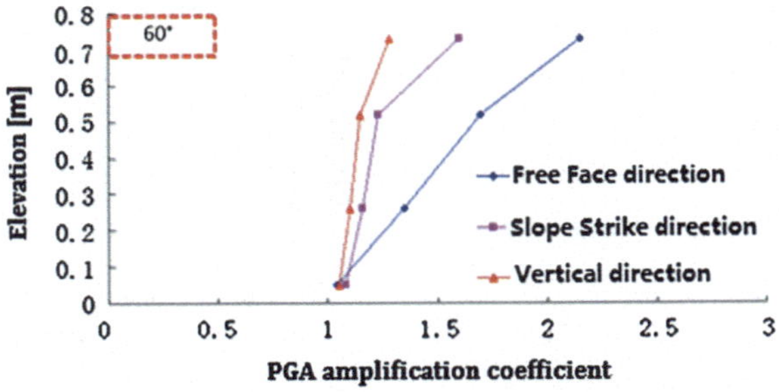

Fig. 3.87 PGA elevation amplification coefficient (PGA = 0.2 g)

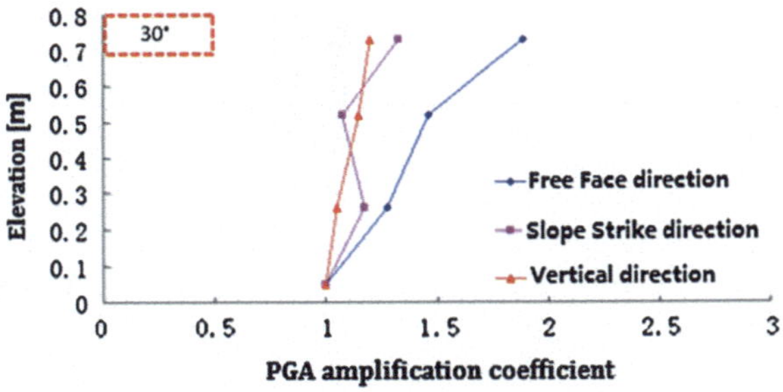

Fig. 3.88 PGA elevation amplification coefficient (PGA = 0.4 g)

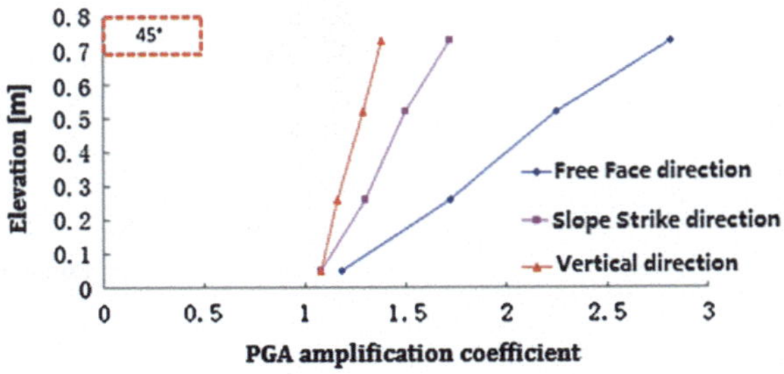

Fig. 3.89 PGA elevation amplification coefficient (PGA = 0.4 g)

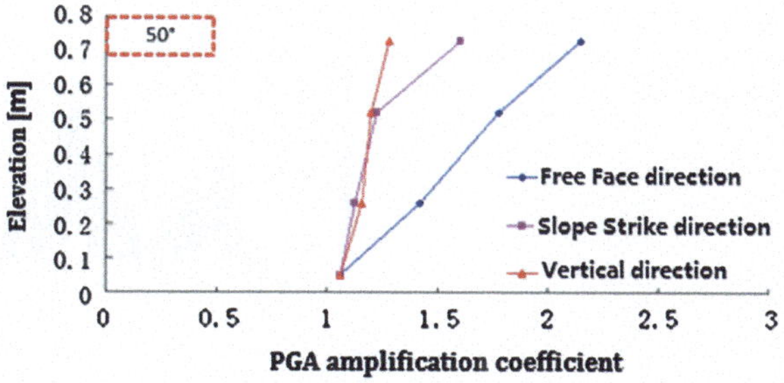

Fig. 3.90 PGA elevation amplification coefficient (PGA = 0.4 g)

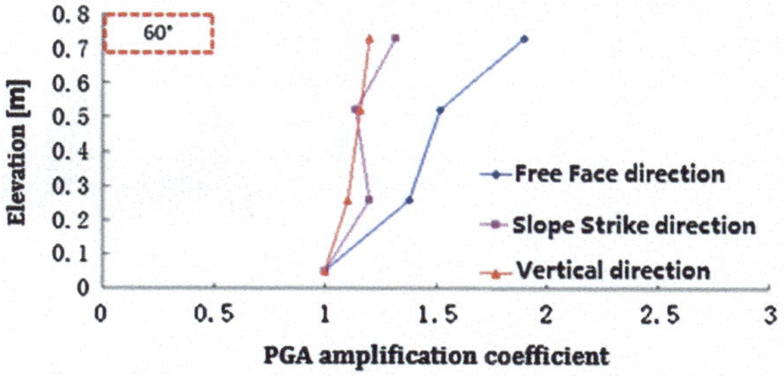

Fig. 3.91 PGA elevation amplification coefficient (PGA = 0.4 g)

3.2.2 Fourier Spectrum and Response Spectrum of Acceleration

According to analysis and summaries of macroscopic seismic damage experience and apparatus observing data, the main characteristics of seismic ground motion, for engineering seismology, also can be described by frequency spectrum and duration besides amplitude. The curves that show the relation between amplitude and frequency or period in an earthquake are collectively known as frequency spectrum, which mainly includes Fourier spectrum and response spectrum. Lots of research results demonstrate that the spectrum components of rock slope seismic ground motion acceleration change with the slope height elevation. Therefore, this shaking table test arranged large amount of monitoring points to study the frequency spectrum characteristic change rules of rock slopes along elevation.

3.2.2.1 Rock Slope Acceleration Fourier Spectrum Change Rules Along Elevation

Due to lack of space, this chapter only selects test points on 45° surface in *L, M,* and *N* three directions under effects of Wenchuan wolong seismic waves with PGA = 0.2 g to analyze. There set up four test points on the 45° surface with respective number of 4#, 10#, 16#, and 22#, and their vertical distances to slope

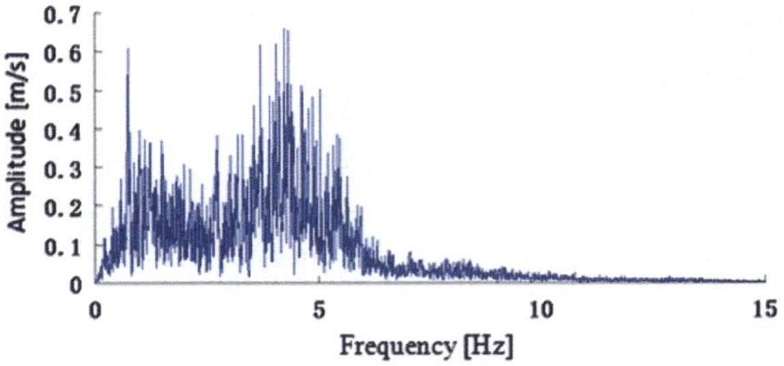

Fig. 3.92 Fourier spectrum of acceleration in free face direction (4#)

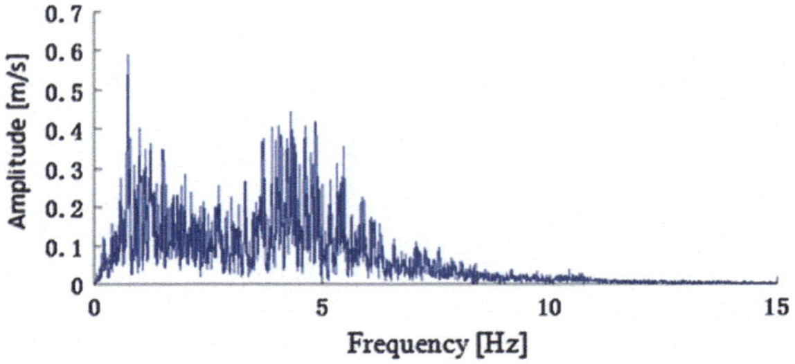

Fig. 3.93 Fourier spectrum of acceleration in free face direction (10#)

foot are, respectively, 0.05, 0.26, 0.52, and 0.73 m. See Figs. 3.92, 3.93, 3.94, 3.95, 3.96, 3.97, 3.98, 3.99, 3.100, 3.101, 3.102, and 3.103 for Fourier spectrum calculation results of all test points.

It is known from Figs. 3.92, 3.93, 3.94, and 3.95 that the dominant frequencies of test points 4#, 10#, 16#, and 22# on 45° surface under effects of Wenchuan wolong seismic waves with PGA = 0.2 g concentrate in the range of 0.0–1.0 Hz and 3.69–5.48 Hz, and with the increase of elevation, the former gradually weakened and the latter gradually strengthened. Figures 3.96, 3.97, 3.98, and 3.99 show that the dominant frequencies of test points 4#, 10#, 16#, and 22# concentrate in the range 3.3–3.6 Hz, and with the increase of elevation, other frequency components gradually weakened. Figures 3.100, 3.101, 3.102, and 3.103 show that the dominant frequencies of test points 4#, 10#, 16#, and 22# in vertical direction concentrate in the range of 4.8–5.0 Hz and 6.1–6.4 Hz, and with the increase of elevation, the former gradually strengthened and the latter gradually weakened. The reason for the above phenomenon lies in that the first natural frequency of the model test model is 4.5 Hz, which enlarges the surrounding frequency components. This is identical with the conclusion of the normally said "the slope soil has filter effects of the high-frequency components of seismic waves and has amplification effects for the low-frequency components. This is because, for normal slope soil, the natural frequency f is small, which has amplification effects for frequency range around $f \pm \Delta f$ in seismic waves and has filter effects for other frequency ranges.

From the above analysis, we know that the changes of Fourier spectrum frequency components of acceleration in free face direction (*L*), slope strike (*M*), and vertical direction (*N*) along elevation are basically identical; i.e., with the increase of elevation, the slope soil has significant amplification effects for frequency

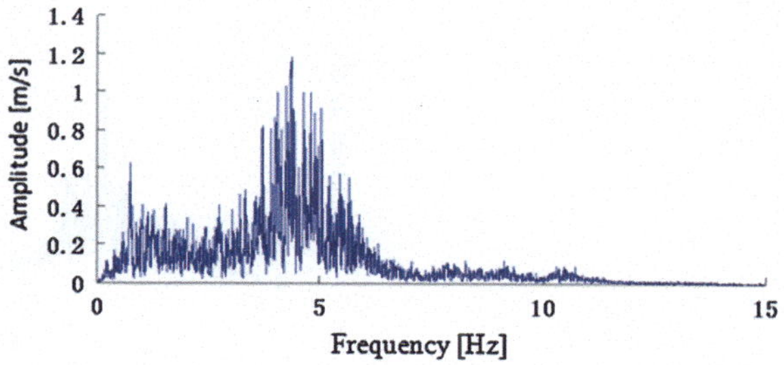

Fig. 3.94 Fourier spectrum of acceleration in free face direction (16#)

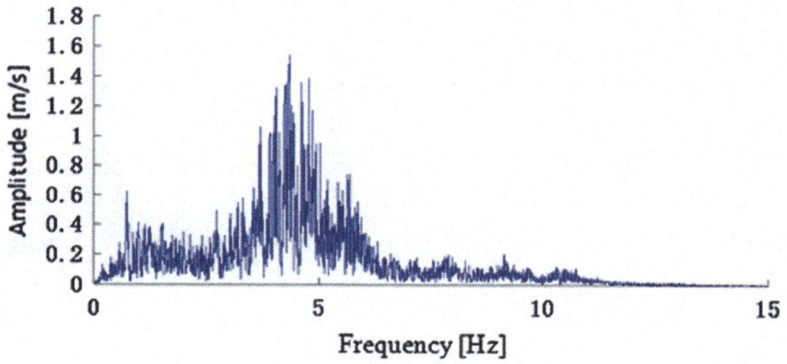

Fig. 3.95 Fourier spectrum of acceleration in free face direction (22#)

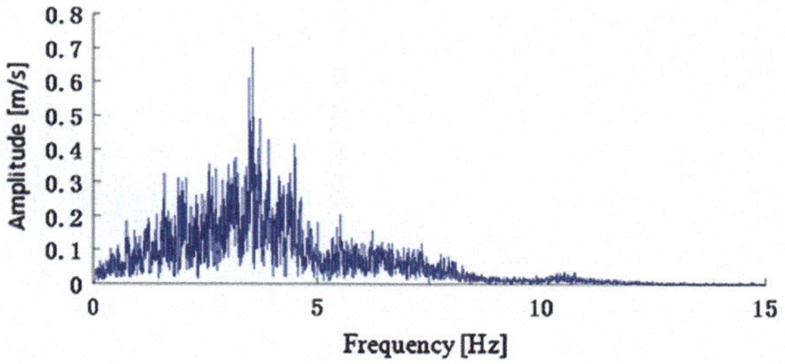

Fig. 3.96 Fourier spectrum of acceleration in slope strike direction (4#)

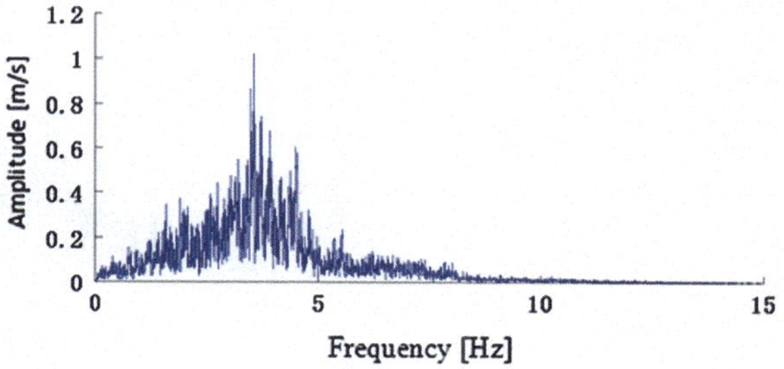

Fig. 3.97 Fourier spectrum of acceleration in slope strike direction (10#)

Fig. 3.98 Fourier spectrum of acceleration in slope strike direction (16#)

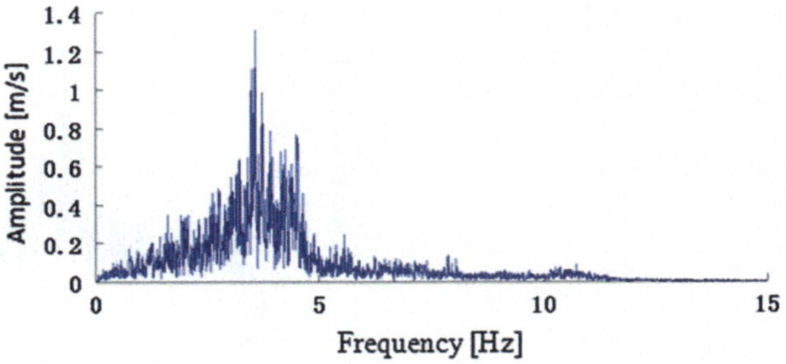

Fig. 3.99 Fourier spectrum of acceleration in slope strike direction (22#)

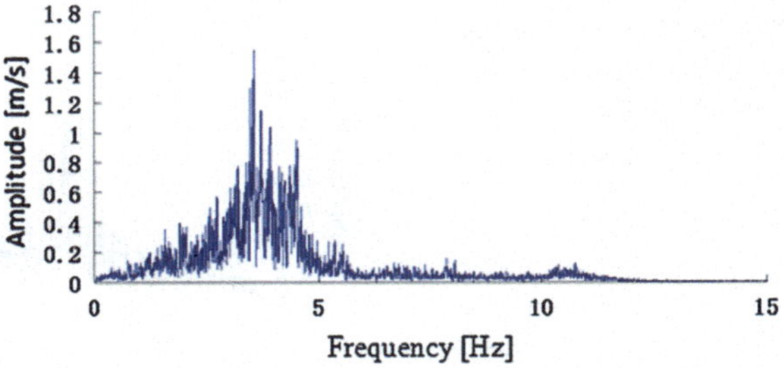

Fig. 3.100 Fourier spectrum of acceleration in vertical direction (4#)

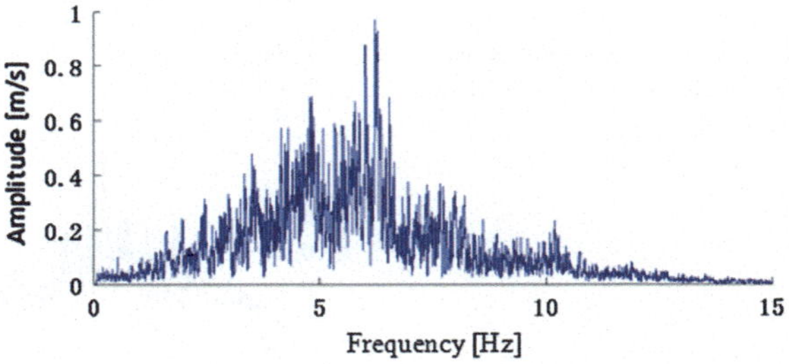

Fig. 3.101 Fourier spectrum of acceleration in vertical direction (10#)

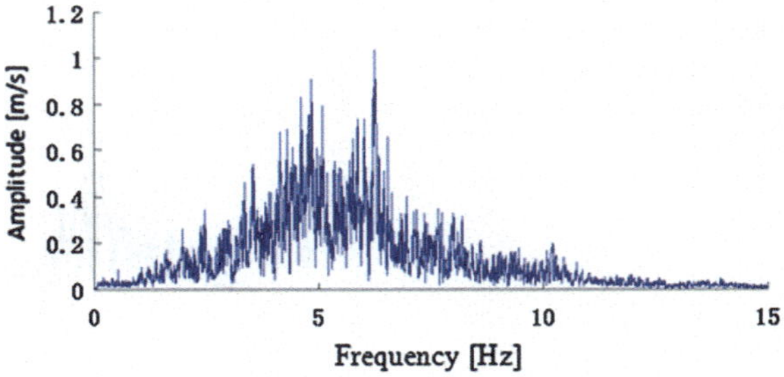

Fig. 3.102 Fourier spectrum of acceleration in vertical direction (16#)

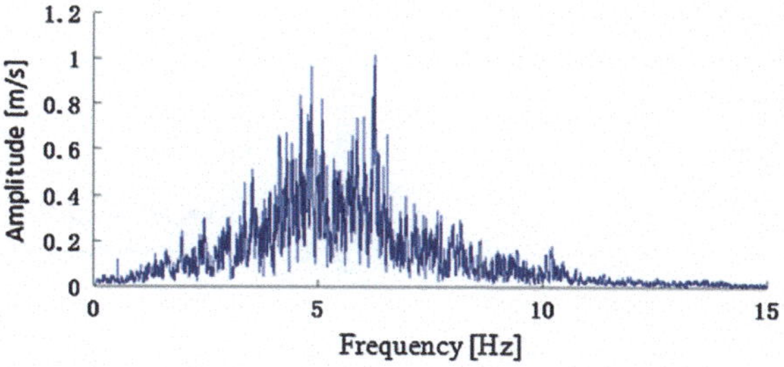

Fig. 3.103 Fourier spectrum of acceleration in vertical direction (22#)

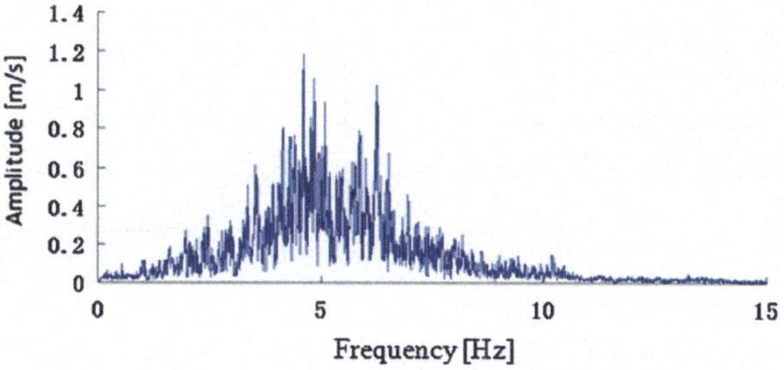

components around its natural frequency range and has filter effects for other frequency ranges.

3.2.2.2 Rock Slope Acceleration Response Spectrum Change Rules Along Elevation

The shape of rock slope acceleration response spectrum has important reference value for recognition of rock slope dynamic response characteristics and for design of rock slope control. In order to study the change rules of rock slope acceleration response spectrum along elevation, this chapter only selects test points on 45° surface in *L, M,* and *N* three directions under effects of Wenchuan wolong seismic waves with PGA = 0.2 g to analyze. There set up four test points on the 45° surface with respective number of 4#, 10#, 16#, and 22#, and their vertical distances to slope foot are, respectively, 0.05, 0.26, 0.52, and 0.73 m. At the same time, in order to more clearly state the elevation amplification effects of different periodic spectrum values in the response spectrum, this chapter takes the three directions' response spectrum of test point 4# at the bottom of the slopes as basis to work out the acceleration response spectrum amplification coefficient of test points 10#, 16#, and 22#. See Figs. 3.104, 3.105, 3.106, 3.107, 3.108, and 3.109 for detailed calculation results. It is noteworthy that this acceleration response spectrum calculation adopts damping ratio of 5 % which is commonly used by engineering.

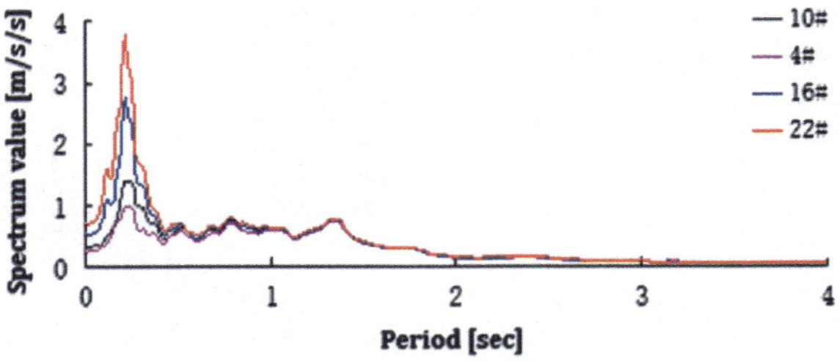

Fig. 3.104 Acceleration response spectrum in free face direction

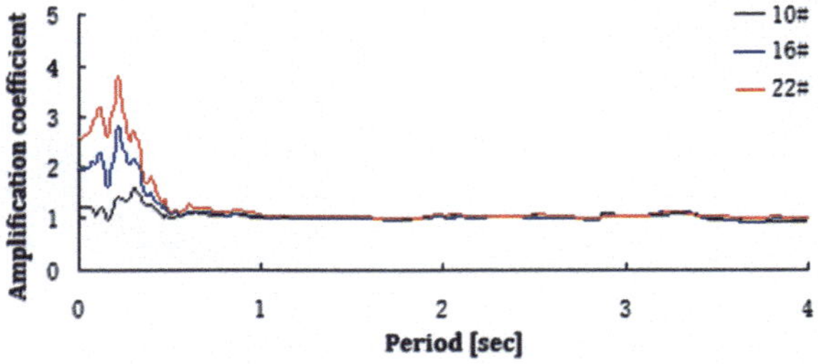

Fig. 3.105 Acceleration response spectrum amplification coefficient in free face direction

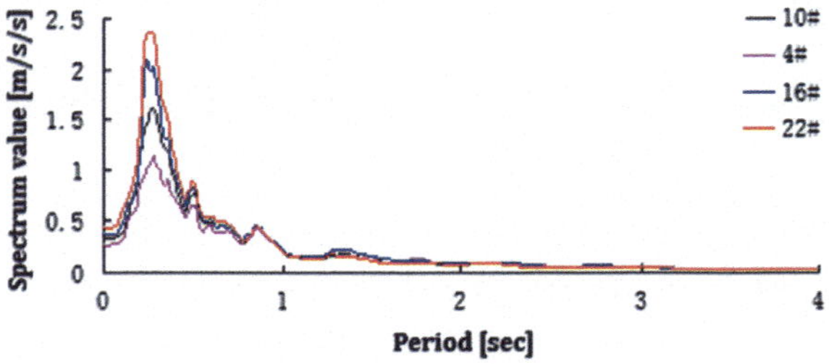

Fig. 3.106 Acceleration response spectrum in slope strike

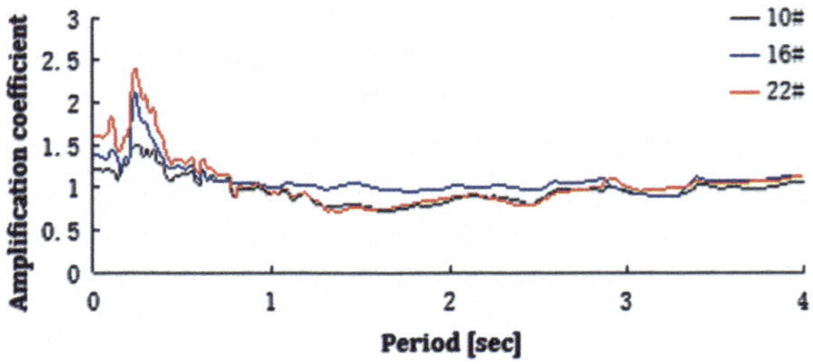

Fig. 3.107 Acceleration response spectrum amplification coefficient in slope strike

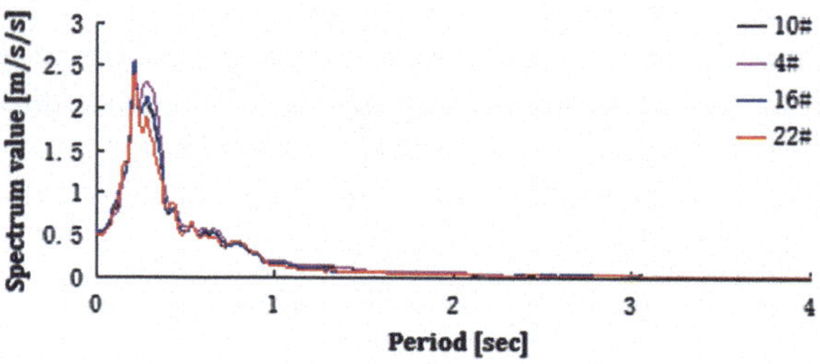

Fig. 3.108 Acceleration response spectrum in vertical direction

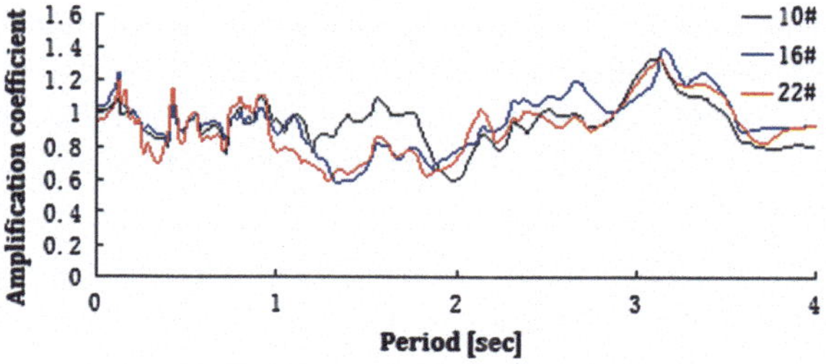

Fig. 3.109 Acceleration response spectrum amplification coefficient in vertical direction

It is known from Figs. 3.104 and 3.105 that the shape of acceleration response spectrum along elevation in free face direction on 45° surface under the effects of Wenchuan wolong seismic waves with PGA = 0.2 g is basically consistent and has peak features. The predominant period concentrates at about $T = 0.22$ s (the corresponding frequency is $f = 4.54$ Hz). The response spectrum amplitude corresponded to short period ($T = 0.0$–0.5 s) has certain amplification effects with amplification coefficient between 1.0 and 4.0, while as for other T periods, especially at long period (low frequency), it has certain reduction effects with amplification coefficient between 0.93 and 1.18. Figures 3.106 and 3.107 show that the shape of acceleration response spectrum along elevation in slope strike direction is basically consistent, and its peak features are not obvious as in free face direction. The predominant period concentrates at about $T = 0.28$ s (the corresponding frequency is $f = 3.57$ Hz). The response spectrum amplitude corresponded to short period ($T = 0.0$–0.8 s) has certain amplification effects with amplification coefficient between 1.0 and 2.5, while as for other T periods, especially at long period (low frequency), it has certain reduction effects with amplification coefficient between 0.76 and 1.05. Figures 3.108 and 3.109 demonstrate that the shape of acceleration response spectrum along elevation in vertical direction has double peak phenomena. The predominant period concentrates at about $T = 0.22$ s (the corresponding frequency is $f = 4.54$ Hz). The response spectrum amplitude corresponded to short period ($T = 0.0$–0.3 s) has slight amplification effects, while as for other T periods, especially at long period (low frequency), the amplification coefficient fluctuates at about 1. The reason for the above phenomenon may lie in that the first natural frequency of the model test model is 4.5 Hz.

It can be known from the above analysis that for the short period, especially around predominant period, the amplification response spectrum in free face direction is stronger than that in slope strike direction, and the amplification response spectrum in vertical direction is the weakest. For the long period, the amplification response spectrum along elevation has certain reduction effects in all three directions.

3.3 Brief Summary

This chapter analyzed the monitored acceleration data under different seismic working conditions, discussed the influence rules of bevel angles, seismic intensity, and seismic ground motion wave types on rock slope acceleration response from the perspective of PGA of slopes, Fourier spectrum and response spectrum and came to the following conclusions:

(1) For mountains containing four slope surfaces, if we analyze the traditional method, i.e., using *X*, *Y*, and *Z* three directions' coordinate system to analyze PGA elevation amplification effects, the results may be unilateral; i.e., *X* direction is the free face direction of one slope surface and slope strike direction of another one, which cannot clearly reflect the influence of bevel angles on PGA elevation amplification effects on each direction. Therefore, when studying the slope dynamic problems, it is suggested to use the three-dimensional local coordinate system, i.e., the free face direction, slope strike direction, and vertical direction of slopes.

(2) With the increase of bevel angles, the PGA elevation amplification effects in vertical direction will rise gradually, while at the same time, there exists a sudden increase turning point at bevel angles of 45° and a levelling off turning point at 50°. However, the PGA elevation amplification effects in slope strikes stay unchanged with the increase of bevel angles, and its steps are comparatively gentle.

(3) With the increase of seismic ground motion PGA, the peak acceleration elevation amplification effect decreases gradually; slope strike direction and vertical direction reduce gradually and show characteristics of magnitude saturation. Under the effects of EL Centro, Wenchuan wolong, and Kobe seismic waves, the rules of peak acceleration elevation amplification effect in free face direction, slope strike direction, and vertical direction are as follows: EL Centro seismic wave > Wenchuan wolong seismic wave > Kobe seismic wave, and free face direction > slope strike direction > vertical direction.

(4) The changes of Fourier spectrum frequency components of acceleration in free face direction (*L*), slope strike (*M*), and vertical direction (*N*) along elevation have certain regularity, i.e., with the increase of elevation, the slope soil has significant amplification effects for frequency components around its natural frequency range *f* and has filter effects for other frequency ranges.

For the short period, especially around predominant period, the amplification response spectrum in free face direction is stronger than that in slope strike direction, and the amplification response spectrum in vertical direction is the weakest. For the long period, the amplification response spectrum along elevation has certain reduction effects in all three directions. In addition, it has obvious peak phenomena for free face direction, while for slope strike direction and vertical direction, the peak phenomenon is less obvious.

Chapter 4
Numerical Analysis Research

Field monitoring results of rock slope seismic arrays and shaking table test results of rock slopes in Xishan Park, Zigong, have revealed the existence of rock slope acceleration elevation amplification effects, and we have come to some valuable conclusions through research on the influence of different aspects of bevel angles, seismic density, seismic wave types and seismic ground motion direction on rock slope seismic ground motion characteristics. While the slopes in Xishan Park, Zigong, and the rock slopes used in the shaking table test are all independent slopes containing four surfaces, in reality there exist large areas of continuous slopes of dozens of meters, hundreds of meters, or even much longer. Thus there must be differences between the dynamic characteristics in reality and the seismic ground motion characteristics of the independent slopes. Therefore, based on the results of previous research, in this chapter we will perform a general analysis of the model and will adopt new numerical calculation software, GDEM, to further analyze the seismic response rules of rock slopes, focusing on the above slope types, to effect a systematic investigation of field array monitoring (macro-phenomena), shaking model test (further research on regularity) and data simulation (micro complement) in order to enrich research content and expand research results.

4.1 Brief Introduction to GDEM

The continuum-based discrete element method, known as CDEM, is a new numerical analysis method developed within the past few years by the Institute of Mechanics of the Chinese Academy of Sciences Joint Laboratory of Discontinuum Mechanics for Engineering Disasters. This method couples the finite element method (FEM) with the discrete element method (DEM), i.e. using FEM inside the slope and DEM at the boundary of the slope. Through cracks of the internal parts and boundaries of slopes, it is able not only to simulate the deformation and motion features of materials in continuous and non-continuous states, but also to obtain the

Y. Changwei et al., *Slope Earthquake Stability*,
DOI 10.1007/978-981-10-2380-4_4

gradual damage course of materials from continuum to non-continuum. The graphics processing unit, or GPU, is the core part of a computer graphics card, and is a natural high performance parallel device. Beijing GDEM Technology Ltd. developed the commercial software, GDEM, with independent intellectual property rights, together with the Institute of Mechanics of the Chinese Academy of Sciences, on the basis of combined CDEM and GPU technology, which improves the computing speed by two orders of magnitude and the computing scale by one order of magnitude.

The GDEM software allows interaction between kernel calculation and model display through hybrid programming of Microsoft Foundation Class (MFC) with OpenGL, and supports graphical user interface (GUI) and command stream dual operation. Its main features are as follows:

First, using the finite volume method (FVM) to work out unit stress and strain, and to enable large rotation calculation through real time updating normal vector and unit area of all unit surfaces;
Second, using a semi-spring model and semi-arris model to work out contact force among units, and converting conclusions on a complex contact state (angle-angle, angle-side, angle-surface, side-side, side-surface and surface-surface contact) into conclusions on a semi-spring and target surface, and semi-arris and target arris.
Third, using explicit calculation technology based on the front insertion method, which can give slope stress and strain features and motion state at any time, and which makes working out ill-conditioned problems and distortion as easy as working out the normal problems.

GDEM software is currently widely used in geotechnical, mining, tunnel, hydraulic, geological and structural engineering, and for gradual damage course simulation and global stability evaluation of geologic materials in all kinds of engineering.

4.1.1 Advantages of GDEM Software

Compared with current numerical software, the advantages of GDEM can be summarized as follows: First, it is suitable for solving dynamic, non-linear and ill-conditioned problems, and there are explicit physical results after each time step; thus there will be no malfunction problems as a result of a stiffness matrix singularity. Second, it will not form the whole stiffness matrix, which saves a lot of memory, so it is suitable for solving large-scale problems. Third, GDEM combines three sets of software [Fast Lagrangian Analysis of Continua (FLAC), Universal Distinct Element Code (UDEC), Particle Flow Code (PFC)], so it can simulate the whole process of geologic calamity body from continuum to non-continuum. Fourth, GDEM uses GPU for acceleration, which is 100–150 times faster than solutions of the same type, and its computing scale can amount to mega-scale grid.

4.1.2 Fundamental Principle of GDEM

Supposing that Ω is a bounded closed domain in a space, and its boundary surface $\partial\Omega$ is composed of limited blocks of smooth or piece-wise smooth surfaces, and supposing that function $f(x, y, z)$ has first-order continuous partial derivatives in Ω, then by the Gauss divergence theorem we can get:

$$\oiint_{\partial\Omega+} f n_i \mathrm{d}S = \iiint_{\Omega} \frac{\partial f}{\partial x_i} \mathrm{d}\Omega \tag{4.1}$$

in which n_i is the unit vector in the outside normal direction of boundary surface.

The average value $\left\langle \frac{\partial f}{\partial x_i} \right\rangle$ of $\frac{\partial f}{\partial x_i}$ in a bounded closed domain can be expressed by Formula (4.2), and V is the total volume of bounded closed domain Ω. When $\Omega \rightarrow 0$, there is $\left\langle \frac{\partial f}{\partial x_i} \right\rangle = \frac{\partial f}{\partial x_i}$

$$\left\langle \frac{\partial f}{\partial x_i} \right\rangle = \frac{1}{V} \iiint \frac{\partial f}{\partial x_i} \mathrm{d}V = \frac{1}{V} \oiint_{\partial\Omega+} f n_i \mathrm{d}S \tag{4.2}$$

Supposing that a polyhedron has N surfaces, then Formula (4.2) can be rewritten as Formula (4.3):

$$\left\langle \frac{\partial f}{\partial x_i} \right\rangle \approx \frac{1}{V} \sum_{k=1}^{N} \bar{f} n_i^k \Delta S^k \tag{4.3}$$

ΔS^k is the area of the kth surface, the classification of outside normal direction of this surface in x_i direction is n_i, and $\bar{f}$ is the average value of f on the kth surface. Taking a tetrahedron as an example, if we want to work out the stress increment inside the unit, we should first work out the velocity gradient average inside the tetrahedron, as Formula (4.4) shows:

$$\left\langle \frac{\partial v_i}{\partial x_j} \right\rangle = \frac{1}{V} \sum_{k=1}^{4} \bar{v_i} n_j^k \Delta S^k \tag{4.4}$$

The stress increment can be expressed as Formula (4.5):

$$\Delta \varepsilon_{ij} = \frac{1}{2} \left(\left\langle \frac{\partial v_i}{\partial x_j} \right\rangle + \left\langle \frac{\partial v_j}{\partial x_i} \right\rangle \right) \Delta t \tag{4.5}$$

The stress increment can be expressed as Formula (4.6):

$$\Delta\sigma_{ij} = C_{ijkl}\Delta\varepsilon_{kl} \tag{4.6}$$

The total component of stress can be expressed as Formula (4.7):

$$\sigma_{ij} = \sum \Delta\sigma_{ij} \tag{4.7}$$

The node stress component of all nodes in the tetrahedron is as Formula (4.8):

$$f_i^N = \sum_{k=1}^{3} \int_s \sigma_{ij} n_j^k \mathrm{d}s \approx \sigma_{ij} \sum_{k=1}^{3} \left(n_j^k \Delta S^k\right) \tag{4.8}$$

Here, N is representative of the Nth node, k is the serial number of the surface containing Node N (each node inside the tetrahedron belongs to three coplanar vectors), n_j^k is the component of a unit vector in the outside normal direction of the surface, and ΔS^k is the node area of Node N in the kth area.

If Node N is shared by M nodes, the calculation of the total joint forces is as shown in Formula (4.9):

$$F_i^N = \sum_{l=1}^{M} \left(f_i^N\right) = \sum_{l=1}^{M} \left[\sigma_{ij} \sum_{k=1}^{3} \left(n_j^k \Delta S^k\right)\right] \tag{4.9}$$

4.1.3 Verification

GDEM uses FVM to work out unit strain and stress, which uses a computing source program and adopts an incremental mode to achieve the above process. Its calculation flow is shown in Fig. 4.1. The dotted line in the figure shows the solution iteration through FVM.

Taking the elastic displacement field calculation of a homogeneous slope as an example to check the correctness of the program, a comparison with the results calculated with FLAC3D is shown in Figs. 4.2 and 4.3. Through comparison, we know that the results of GEDM calculated based on FVM are exactly the same as those with FLAC3D.

Using the incremental method, the classic ideal elastoplastic criterion—the Drucker-Prager and Mohr-Coulomb criteria—is introduced. With the uniaxial compression test, the results calculated by GEDM software are identical to those of the commercial FLAC3D software (see Figs. 4.4 and 4.5).

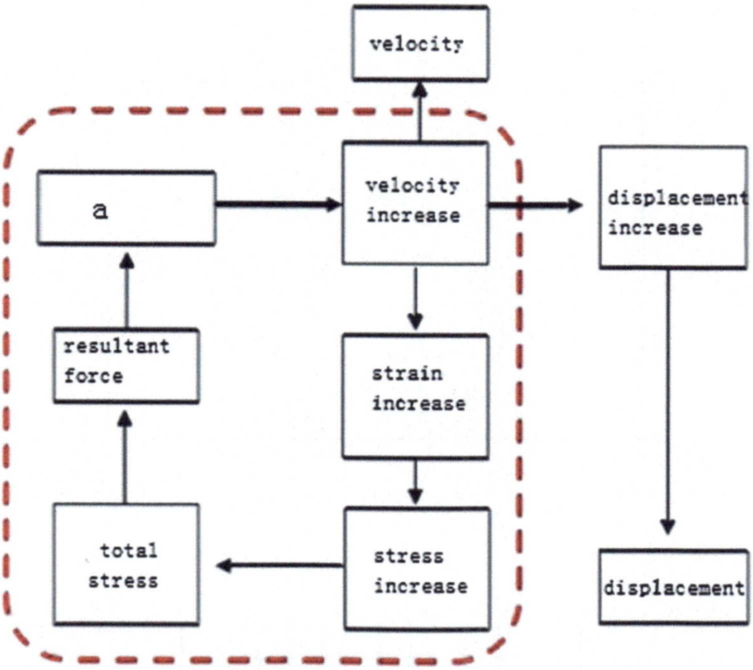

Fig. 4.1 Calculation flow of FVM

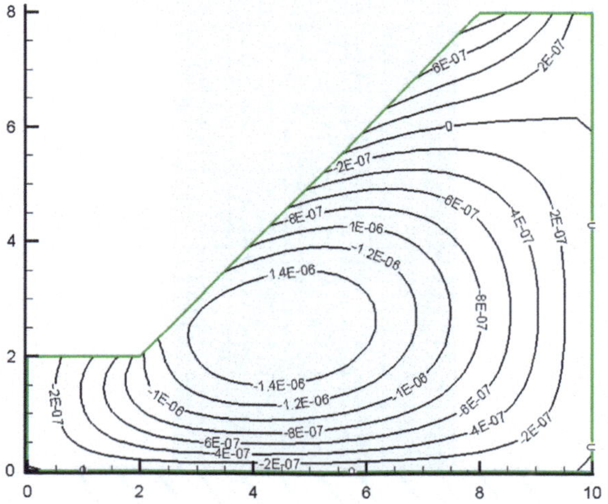

Fig. 4.2 Rock slope horizontal displacement field comparison diagram based on GEDM

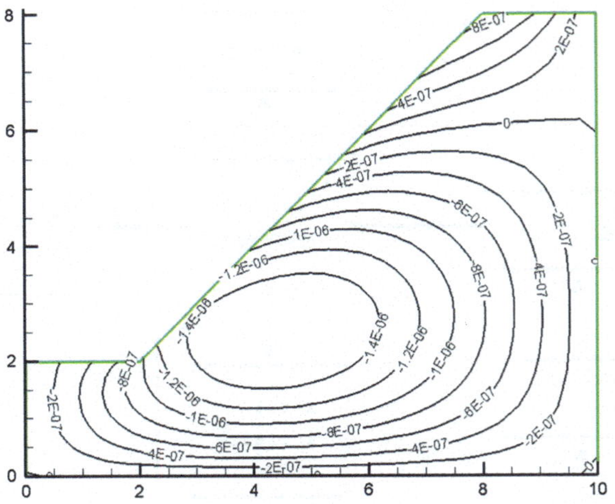

Fig. 4.3 Rock slope horizontal displacement field comparison diagram based on FLAC3D

Fig. 4.4 Model diagram of uniaxial compression test

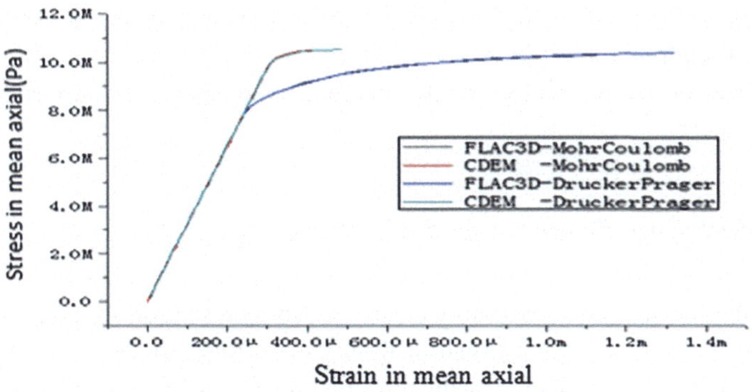

Fig. 4.5 Results comparison diagram of uniaxial compression test

4.1.4 Fundamental Function

The fundamental function of GDEM mainly includes a pre-processing function, a main body function and a post-processing function. The main contents are as follows:

(1) Pre-processing function
GDEM can produce a simple two-dimensional rectangle and homogeneous slope model network. The software supports mesh import of commercial software such as ANSYS, Patran and FLAC3D.

(2) Main body function

① Block models: linear elastic model, transversely isotropic model, ideal elastoplastic model (Drucker-Prager and Mohr-Coulomb models), strain-softening model based on the Mohr-Coulomb criterion, ubiquitous joint model based on the Mohr-Coulomb criterion, creep model (coupling of Burger with Mohr-Coulomb and Kelvin plasticity model), excavation model and pore percolation model.

② Joint models: linear elastic model, brittle fracture model, strain-softening model and ideal elastoplastic model.

③ Bar models: pile model, beam model, bolt model and anchor cable model.

④ Kinetics-based models: explosion loading, seismic loading, the applied module to impact loading, non-reflecting boundary conditions, and the applied module under free field boundary conditions.

(3) Post-processing function

① Checks the acceleration, velocity, displacement, stress, strain, flow and flow velocity nephogram of the rod model and real units.

② Checks the time-history curve of monitoring points.

③ Draws real unit velocity graph, displacement vector graph and principal stress direction graph.
④ Draws the unit displacement, force and moment vector graphs of the rod model.

4.1.5 Modeling Progress and Methods

Before modeling, we should begin by clearly recognizing the subordinate system of GDEM and build a two-dimensional or three-dimensional model according to our needs. The software's default straight up direction is the *y*-axis, which is also the positive direction, and the opposite is the negative direction. We used to call the due north direction the software's *z*-axis, which is also the positive direction, and the opposite is the negative direction. We have a tendency to conform to the geographical location with due north at 0°, and the dip angle is identical to the recognized principle in geology. The *x*-axis takes due east as the positive direction and the opposite as the negative direction, which suits geographic orientation.

We mentioned previously that in the pre-processing function, GDEM can generate a simple two-dimensional model on its own, but for complex models it needs to import from other software. GDEM can operate with both GUI and a command stream, and it provides rich commands for simple and easy modeling and adjustment processes. It largely adopts the three modes:

(1) Using GUI operation;
(2) Using single command input;
(3) For complex modeling needing a large number of commands, it can use text input and directly modify and save over the texts.

A complete joint model needs six steps in sequence to finish. The software should: (1) produce a simple two-dimensional rectangle and homogeneous slope model network, or support mesh import of commercial software such as ANSYS, Patran and FLAC3D; (2) input parameters such as dip angle, inclination, initial point, quantity, spacing and continuity of joints; (3) set up material parameters for the model as well as the mechanical parameters of the joint itself; (4) set up monitoring curve or stress nephogram of displacement, velocity, stress and strain, according to the user's needs; (5) set up boundary conditions of the model; and (6) work out the results and output it.

4.2 Key Problems in Numerical Analysis

In the general case, continuous slopes of dozens of meters, hundreds of meters, or even much longer can be simplified into plane strain problems. Regarding the influence of factors such as the nature of rock and earth mass and the seismic wave types, the above-mentioned rock slopes can be generalized into two major types:

single-sided rock slope and double-sided rock slope. Neither field seismic array monitoring nor shaking table tests have been conducted on these slopes. Therefore, this chapter will focus on these two types of rock slopes, select a classic hazard point in the 5.12 Wenchuan earthquake, and use the shaking table test combined with numerical simulation to study seismic ground motion characteristics of rock slopes under the effects of earthquakes from a general rule in order to expand the research results. What is important to note is that the shaking table test in this chapter was completed by both our research group and the research group led by Cui Peng from the Chengdu Institute of Mountain Hazards and Environment of the Ministry of Water Resources, Chinese Academy of Sciences, and most of the research results were used by the latter. In addition, because of the external conditions, limited monitoring points were set up for the shaking table test, so it was unable to comprehensively reflect the seismic ground motion characteristics of rock slopes. Therefore, this chapter will only build a numerical analysis model based on the shaking table test model and select parts of the shaking table test data to verify the correctness of the numerical analysis results.

4.2.1 Numerical Analysis Model

The numerical analysis model in this chapter is built by a geometric similarity ratio of 1:1 according to the shaking table test model, whereas the experimental model takes the prototype of two classic landslide hazard points by the left side of National Highway G213, and is built using a geometric similarity ratio of 1:600. Its geometric features are as follows: the single-sided high and steep model is 1807 mm high, 3500 mm wide, and its bevel angle is in the range of 50°–60°; the river valley at the slope foot is 173 mm, and the riverbed is slightly inclined; the bevel angle of the river valley's right bank slope is in the range of 30°–40°, and the slope top is partly covered by a seriously weathered layer. See Fig. 4.6 for the concrete experiment model. The double-sided high and steep model is 1810 mm high and 3500 mm wide, which includes steep slope topography and gentle slope topography with bevel angles in the range of 50°–60° and 42°–50°, respectively. At the same time, in order to more truly reflect the practical situations of the above-mentioned slopes, the soil layer and weathered layer on the surface of the slopes are generalized into slope wash in slope top, mid-slope and slope foot according to field survey results. See Fig. 4.7 for the detailed shaking table test model. In addition, to thoroughly investigate the seismic ground motion characteristics of high and steep rock slopes, we set up many monitoring points inside and on the surface of the numerical model to monitor the dynamic response time-history changes of slope acceleration, stress and strain under seismic effects. Among them, we select monitoring points K8, K11, K14 and B6, B8, B4 as comparison monitoring points of the single-sided and double-sided rock slopes to verify the correctness of the numerical results. See Figs. 4.8 and 4.9 for numerical analysis model and monitoring point distribution of the single-sided and double-sided rock slopes.

Fig. 4.6 Shaking table test model of single-sided rock slope

Fig. 4.7 Shaking table test model of double-sided rock slope

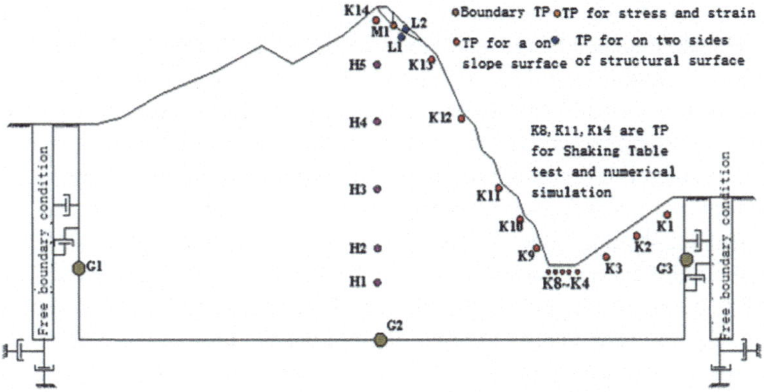

Fig. 4.8 Numerical model of single-sided rock slope

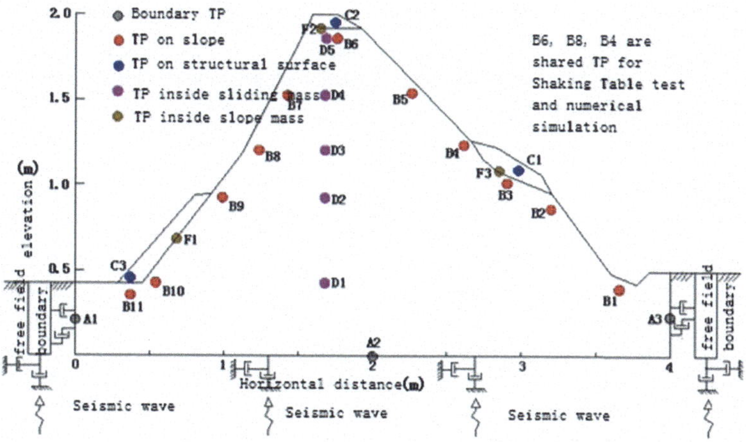

Fig. 4.9 Numerical model of double-sided rock slope

4.2.2 Material Parameters and Constitutive Model

Based on the physico-mechanical parameters of the slope materials of the prototype, the above-mentioned shaking table test used quartz sand, barite powder, clay, plaster or fly ash and water at different ratios to build the model in order to simulate dolomitic limestone of the single- and double-sided high and steep slopes and surface weathering rock mass materials of the cover layer. The numerical simulation of this part adopted materials identical to those in the shaking table test, and the concrete physico-mechanical parameters can be seen in Table 4.1. The sliding bed and slip mass adopted an elastoplastic model and Mohr-Coulomb (M-C) failure principle, while the structural surface and the particles inside the slip mass used a brittle fracture constitutive model and M-C failure principle.

4.2.3 Boundary Conditions

When carrying out dynamic numerical simulation of single-sided and double-sided high and steep slopes, it is important to exert reasonable artificial boundaries. Boundaries are set up according to the following steps: ① constrain the bottom boundary in a vertical direction and the bilateral boundaries in a horizontal direction; and ② change the original model displacement boundary conditions into free field boundaries and the bottom boundary constraints in a vertical direction into viscid boundaries so as to reduce energy loss due to seismic wave reflection.

Table 4.1 The physico-mechanical parameters of the shaking table test model materials

Model name	Unit weight (kN/m^3)	Cohesive force (kPa)	Internal friction angle (°)	Elastoplastic model (MPa)	Poisson ratio
Single-sided slope	23	13.7	39.4	100	0.2
Double-sided slope	22	15.6	37.9	75	0.25
Cover layer	21	8.3	28.8	10	0.35
	Structural position	Normal stiffness (MPa)	Shear stiffness (MPa)	Internal friction angle (°)	Cohesive force (kPa)
Single- and double-sided slopes	Inside of cover layer	600	300	32	8.1
	Bedrock cover layer	4500	2300	32	2.3

4.2.4 Seismic Ground Motion Input Conditions

The seismic wave selected in the shaking table test is the adjusted Wenchuan Wolong seismic wave by a time-simulation ratio of 1:4, and in this chapter we select the same seismic wave to conduct numerical analysis. The real seismic wave peak ground acceleration (PGA) of the Wenchuan Wolong seismic array is 957 and 857 cm/s^2 in the horizontal and vertical directions, respectively, and the duration is 163 s. After adjusting by the time-simulation ratio, the seismic wave duration is 51 s, and the amplitude is unchanged. See Figs. 4.10 and 4.11 for detailed acceleration time-history curves.

During the GDEM numerical analysis process, the dynamic loading represented by the acceleration time-history curve cannot be exerted directly on the viscid boundaries, so it must be converted into a stress curve before exerting on the bottom boundary. This chapter will use numerical integration to convert the acceleration

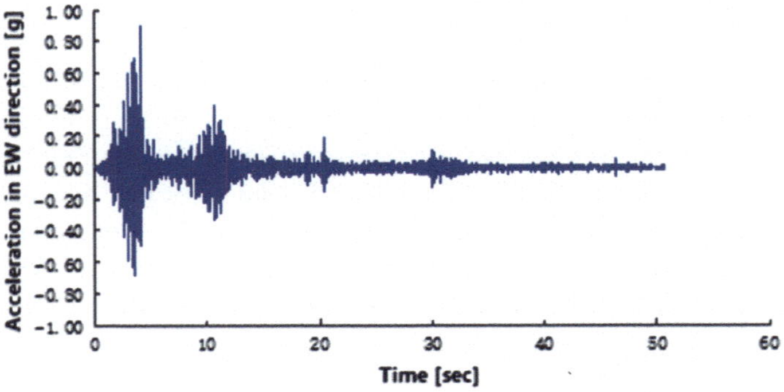

Fig. 4.10 Acceleration time history in EW direction (0.9 g)

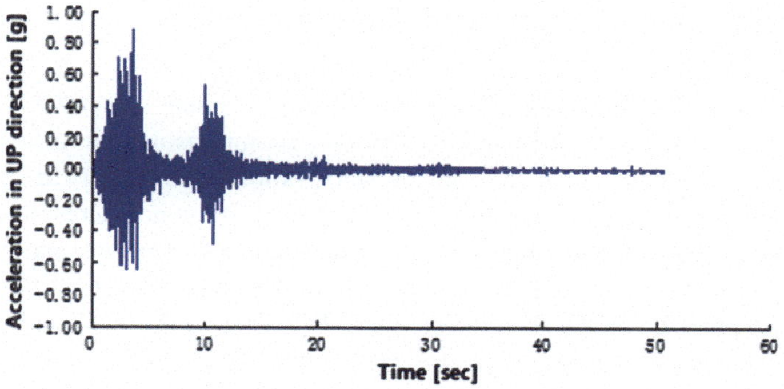

Fig. 4.11 Acceleration time history in UP direction (0.9 g)

time-history curve into a velocity time-history curve, and will then use Formulas (4.10)–(4.13) to convert the velocity time-history curve into a stress time-history curve and exert on the bottom boundary:

$$\sigma_{\mathrm{n}} = -2(\rho C_{\mathrm{p}})\upsilon_{\mathrm{n}}; \tag{4.10}$$

$$\sigma_{\mathrm{S}} = -2(\rho C_{\mathrm{S}})\upsilon_{\mathrm{S}} \tag{4.11}$$

$$C_{\mathrm{P}} = \sqrt{\frac{K + 4G/3}{\rho}}; \tag{4.12}$$

$$C_{\mathrm{S}} = \sqrt{G/\rho} \tag{4.13}$$

In the formula, σ_{n} is the exerted normal stress (MPa), σ_{S} is the exerted tangential stress (MPa), ρ is medium density (kg/m^3), C_{P} is the P wave velocity of medium (m/s), C_{S} is the S wave velocity of medium, υ_{n} is the vibration speed of the input mass point in the vertical direction, υ_{S} is the vibration speed of the input mass point in the horizontal direction, K is the medium bulk modulus, and G is the medium shear modulus.

According to suggestions by Kuhlemeyer and Lysme [149], the grid size is controlled by the shortest wavelength of input wave. Supposing that the largest size of the grid is Δl and the shortest wavelength of the input seismic wave is λ, then Δl must be less than (1/10–1/8) λ. Supposing that the highest frequency of the input wave that does not distort the wave shape is $f_{\max}$, then we can calculate according to Formula (4.14):

$$f_{\max} = \frac{C}{\lambda} = \frac{C}{10\Delta l} \tag{4.14}$$

Here, C has a small longitudinal and transverse wave velocity.

4.3 Verification of Numerical Analysis Results

In order to verify the correctness of the numerical calculation method, we set up monitoring points K8, K11, K14 and B6, B8, B4 as comparison monitoring points in single-sided and double-sided rock slopes, and compare the numerical calculation results with the shaking table test results. See Table 4.2 for concrete results. To more clearly explain the difference between the shaking table test results and numerical analysis results in time history, we will list the horizontal and vertical time-history curve of test point B6 in T = 0–4 s (including PGA) (see Figs. 4.12 and 4.13).

Comprehensively analyzing Figs. 4.2, 4.12 and 4.13, we know that the maximum error of horizontal acceleration and vertical acceleration of test points K8, K11, K14 and B4, B6, B8 are all less than 10.0 %, and the minimum error of horizontal acceleration amounts to −1.19 %, while the minimum error of vertical acceleration amounts to 2.28 %. Therefore, the numerical simulation results are basically identical to those of the shaking model test, which fully shows that the numerical model of this dissertation is correct and the calculation results are credible.

Table 4.2 PGA results comparison of shaking table test and numerical simulation

Monitoring point no.	GDEM calculation results (m/s/s)		Shaking table test results (m/s/s)		Error (%)	
	Horizontal acceleration	Vertical acceleration	Horizontal acceleration	Vertical acceleration	Horizontal acceleration	Vertical acceleration
K8	16.6	13.1	16.8	12.8	−1.19	2.34
K11	24.6	31.2	24.0	29.9	2.50	4.35
K14	19.3	15.6	20.5	14.3	−6.21	−8.3
B4	24.9	21.4	23.8	22.3	4.62	9.71
B6	27.5	32.5	27.0	30.5	1.75	6.54
B8	23.2	31.2	24.6	30.5	−5.69	2.28

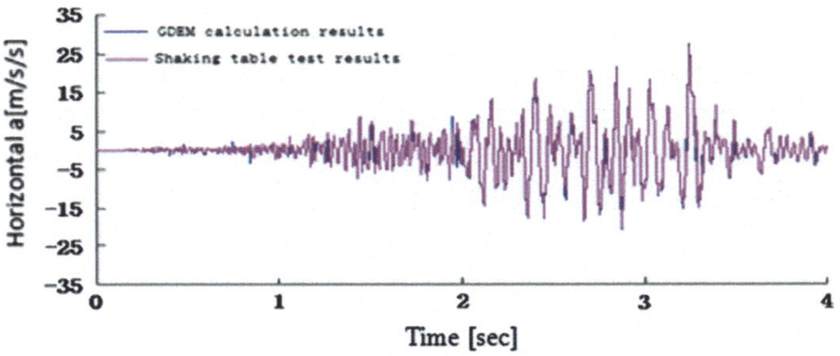

Fig. 4.12 Horizontal time-history curve of test point B6

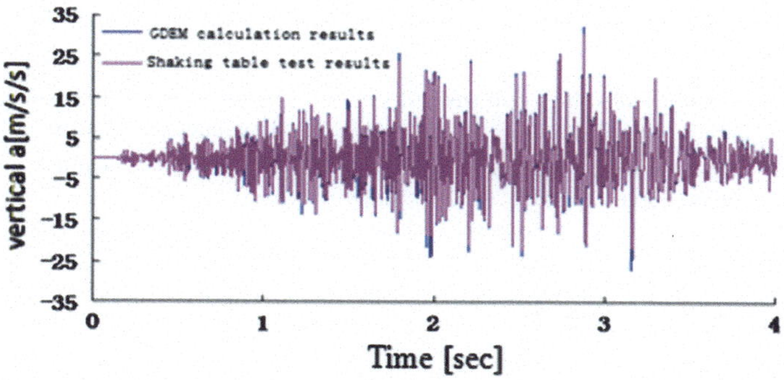

Fig. 4.13 Vertical time-history curve of test point B6

4.4 Seismic Dynamic Response Features

Based on the numerical analysis results of single-sided and double-sided high and steep slopes, we will reveal the general rules of high and steep rock slope dynamic response features from the perspective of time domain and frequency.

4.4.1 Acceleration Amplification Effects

In order to study the propagation rule of acceleration along a slope surface and slope mass elevations in single- and double-sided high and steep slopes under seismic effects, we set up test points B1–B6 (except B3) in gentle slope topography of double-sided high and steep slopes, B6–B10 in steep slope topography, and D1–D5 inside the slope mass. At the same time, we set up test points K14–K10 on the slope surfaces of single-sided high and steep slopes, and H2–H5 and K14 inside the slope mass (see Tables 4.3 and 4.4 for detailed results), and we draw Figs. 4.14 and 4.15 according to data from Tables 4.14 and 4.15.

Table 4.3 Horizontal PGA amplification coefficient of single- and double-sided high and steep slopes

Height to the model bottom (m)	Double-sided high and steep slopes			Height to the model bottom (m)	Single-sided high and steep slopes	
	Steep slope topography	Gentle slope topography	Inside slope mass		On the slope surface	Inside slope mass
1.75	1.91	1.91	1.93	1.78	1.84	1.84
1.35	1.81	1.73	1.65	1.53	2.23	2.15
1.05	1.73	1.61	1.51	1.21	2.72	2.05
0.75	1.62	1.52	1.32	0.84	2.35	1.87
0.25	1.52	1.34	1.15	0.32	1.57	1.17

Table 4.4 Vertical PGA amplification coefficient of single- and double-sided high and steep slopes

Height to the model bottom (m)	Double-sided high and steep slopes			Height to the model bottom (m)	Single-sided high and steep slopes	
	Steep slope topography	Gentle slope topography	Inside slope mass		On the slope surface	Inside slope mass
1.75	1.36	1.36	1.35	1.78	1.06	1.06
1.35	1.35	1.22	1.08	1.53	1.95	1.57
1.05	1.30	0.90	0.84	1.21	2.13	1.91
0.75	1.21	0.80	0.73	0.84	2.12	1.81
0.25	1.10	0.68	0.65	0.32	0.89	0.87

Note The acceleration coefficient in Tables 4.3 and 4.4 is equal to the peak acceleration calculation results/peak acceleration in valley bottom

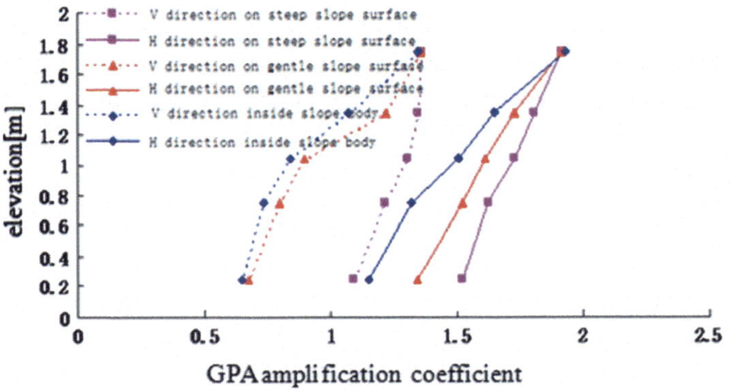

Fig. 4.14 Acceleration amplification effects distribution map of double-sided high and steep slopes

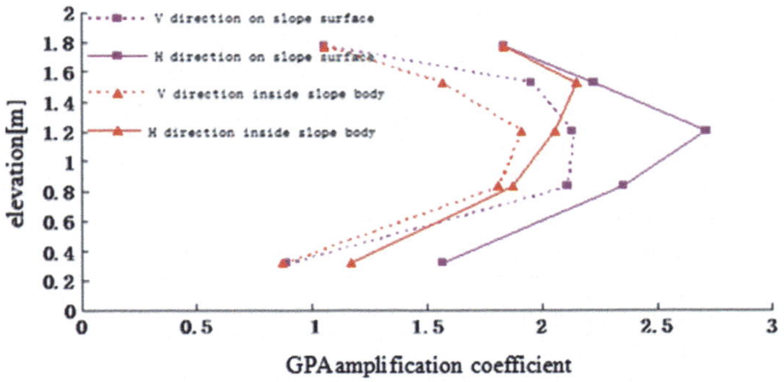

Fig. 4.15 Acceleration amplification effects distribution map of single-sided high and steep slopes

Comprehensively analyzing Tables 4.3, 4.4 and Tables 4.14 to 4.15, we know that for double-sided high and steep slopes, taking acceleration in the valley bottom as the criterion, the horizontal acceleration coefficient on steep slope topography is between 1.52 and 1.81, and the vertical acceleration coefficient is between 1.10 and 1.36. The horizontal acceleration coefficient on gentle slope topography is between 1.34 and 1.91, and the vertical acceleration coefficient is between 0.68 and 1.36. The horizontal acceleration coefficient inside the slope mass is between 1.15 and 1.93, and the vertical acceleration coefficient is between 0.65 and 1.35. For single-sided high and steep slopes, the horizontal acceleration coefficient on the slope surface is between 1.57 and 2.72, and the vertical acceleration coefficient is between 0.89 and 2.13. The horizontal acceleration coefficient inside the slope mass is between 1.17 and 2.15, and the vertical acceleration coefficient is between 0.87 and 1.91. Therefore, whether for a single-sided high and steep slope or for a double-sided high and steep slope, the PGA of different positions along elevations all amplify to different extents, as follows: the amplification effects of a double-sided high and steep slope on acceleration is greater than the amplification effects of single-sided high and steep slope on acceleration, which fully explains the hazard phenomenon of the 5.12 Wenchuan earthquake, where the damage of double-sided slopes or independent slopes is more serious than that of single-sided slopes. PGA amplification effects on the slope surface are greater than PGA amplification effects inside the slope mass, which explains the clear "amplification effect in the free face direction" of the high and steep slopes. The horizontal acceleration coefficient is greater than the vertical acceleration coefficient *i*, and the acceleration coefficient on steep slope topography is greater than the acceleration coefficient on gentle slope topography, which is the same as the shaking table test results, and explains landslide hazard distribution directivity in the Wenchuan earthquake, i.e. the seismic damage on steep slope topography is more serious than that on gentle slope topography.

4.4.2 Acceleration Fourier Spectrum

In order to study the distribution rules of the surface acceleration Fourier spectrum along elevations on single- and double-sided high and steep slopes under seismic effects, we set up test points B1, B4, B6 and B11, B8, B6, respectively, on gentle slope topography and steep slope topography on double-sided high and steep slopes, and at the same time sets up test points K8, K11 and K14 on a single-sided high and steep slope surface. See Fig. 4.16 for detailed results. It is noteworthy that (a)–(i) in Fig. 4.16 represent the calculation results of horizontal acceleration Fourier spectrum along elevations, and (A)–(I) represent the calculation results of vertical acceleration Fourier spectrum along elevations.

Through comprehensive analysis of Fig. 4.16, we know that for both single-sided and double-sided high and steep slopes, the horizontal and vertical acceleration Fourier spectra on slope surfaces along elevations all show the

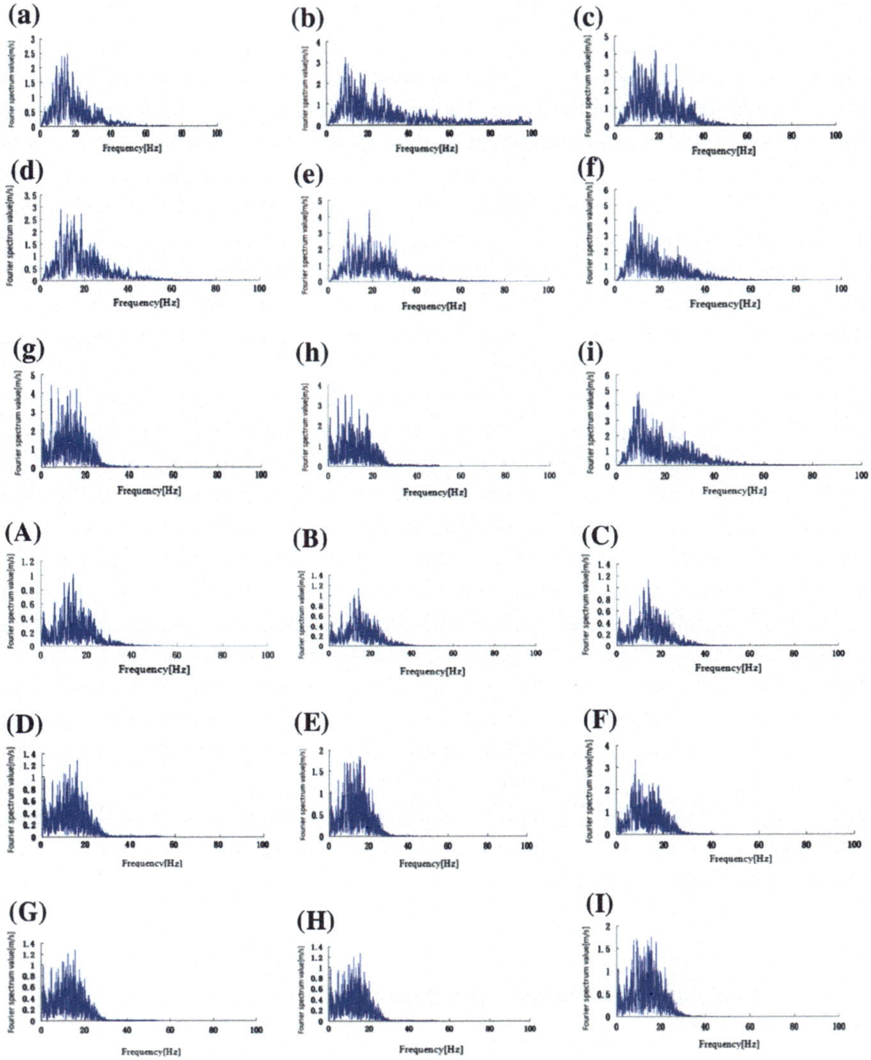

Fig. 4.16 **a** Acceleration Fourier spectrum of test point K8, **b** acceleration Fourier spectrum of test point K11, **c** acceleration Fourier spectrum of test point K14, **d** acceleration Fourier spectrum of test point B1, **e** acceleration Fourier spectrum of test point B4, **f** acceleration Fourier spectrum of test point B6, **g** acceleration Fourier spectrum of test point B11, **h** acceleration Fourier spectrum of test point B8, **i** acceleration Fourier spectrum of test point B6. **A** Acceleration Fourier spectrum of test point K8, **B** acceleration Fourier spectrum of test point K11, **C** acceleration Fourier spectrum of test point K14, **D** acceleration Fourier spectrum of test point B1, **E** acceleration Fourier spectrum of test point B4, **F** acceleration Fourier spectrum of test point B6, **G** acceleration Fourier spectrum of test point B11, **H** acceleration Fourier spectrum of test point B8, **I** acceleration Fourier spectrum of test point B6

changing rule of low frequency component amplification. This chapter will take the horizontal and vertical acceleration on single-sided high and steep slope surfaces as an example to provide an explanation. Figure 4.16 shows that the dominant frequencies of horizontal acceleration at the slope foot, mid-slope and slope top of single-sided high and steep slopes are concentrated in the range of 5–10 Hz, in which test points K8, K11 and K14 show that the spectral value of horizontal acceleration dominant frequencies are 2.51, 3.24 and 4.16, and those of vertical acceleration dominant frequencies are 1.02, 1.14 and 1.25, respectively. Therefore, the horizontal and vertical acceleration Fourier spectra on slope surfaces of single-sided high and steep slopes along elevations all show the changing rule of low frequency component amplification. The above phenomenon may be caused by material damping of the soil mass absorbing some energy of the seismic waves. This has filter effects for seismic waves of high frequency and amplification effects for the section around the first step natural frequency (about 6.06 Hz), which is consistent with the results of the shaking table test and actual field seismic array measurements.

At the same time, to further study the low frequency section amplification rule of the acceleration Fourier spectrum along surface elevations, this chapter selects the peak value of the low frequency section of single- and double-sided high and steep slope surfaces, and draws the peak value changes of the spectrum along elevations in the low frequency section (see Fig. 4.17). The figure shows that regardless of whether for single- or double-sided high and steep slopes, the peak values of the acceleration Fourier spectrum along elevations at different positions all amplify to a different extent. This in turn shows that the amplification effects of double-sided high and steep slopes on the acceleration Fourier spectrum are greater than the amplification effects of single-sided high and steep slopes, the horizontal amplification effects of the acceleration Fourier spectrum are greater than the vertical amplification effects, and the amplification effects of the acceleration Fourier spectrum on steep slope surfaces (with bevel angles of 50°–60°) are greater than those on gentle surfaces (with bevel angles of 42°–50°). Through comprehensive analysis of Figs. 4.16 and 4.17, we know that the slope topography and bevel angles have a significant influence on the changing rules of low frequency component amplification.

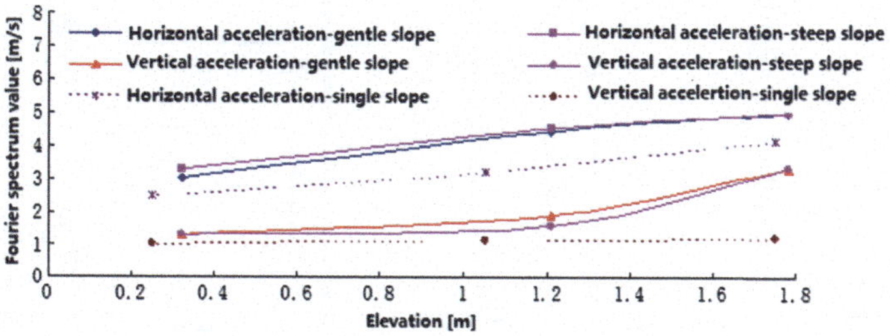

Fig. 4.17 The amplification coefficient of Fourier spectrum peak value in the low frequency section along elevations

4.4.3 Acceleration Amplification Effects

In order to study the distribution rules of the surface acceleration response spectrum along elevations on single- and double-sided high and steep slopes under seismic effects, we set up test points K8, K11 and K14 on single-sided high and steep slope surfaces and test points B1, B4, B6 and B11, B8, B6 on gentle slope and steep slope topography on double-sided high and steep slopes, respectively (see Fig. 4.18 for detailed results). Note that the red frame shows the horizontal and vertical acceleration response spectrum and the blue frame shows the horizontal and vertical acceleration response spectrum value amplification coefficient along elevations.

We know from Fig. 4.18 that the acceleration response spectrum and spectral ratio of single- and double-sided high and steep slope surfaces have a large difference. First, for both horizontal acceleration and vertical acceleration, the acceleration response spectrum of single-sided high and steep slopes along elevations has a single peak value, while the acceleration response spectrum of double-sided high and steep slopes along elevations is a process changing from a single peak value to a double peak value, i.e. the acceleration spectrum at the slope foot has a single peak value, at mid-slope has a slight double peak value, and at the slope top has an obvious double peak value. The reason for this phenomenon may be that the acceleration of double-sided high and steep slopes gradually amplifies along elevations under seismic effects, and the ground motion is more intense, so a whiplash effect arises and results in a double peak value. Second, the maximum horizontal acceleration response spectral value amplification coefficients of single- and double-sided high and steep slopes all occur at period TH = 0.16 s, which is close

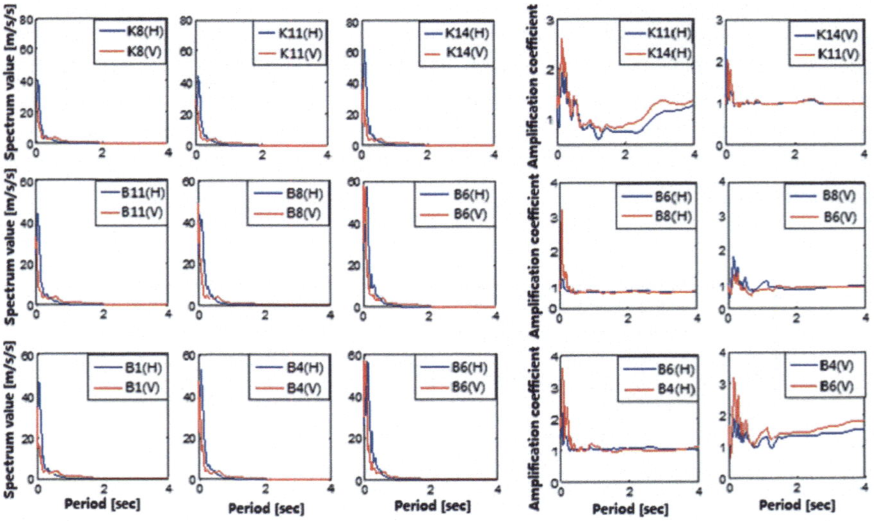

Fig. 4.18 Acceleration response spectrum and spectral ratio of single- and double-sided high and steep slope surfaces

to the natural frequency period $T = 0.165$ s of single-sided high and steep slopes and $T = 0.156$ s of double-sided high and steep slopes, resulting in resonance and causing intense seismic ground motion response on their surfaces. Lastly, around a short period ($T = 0.0$–0.8 s), the horizontal and vertical acceleration response spectral value amplification coefficients of single-sided high and steep slopes are in the range of 1.0–2.5 and 1.0–2.1, that of gentle slope surfaces are in the range of 1.0–3.2 and 1.0–1.82, and those of steep slope surfaces are in the range of 1.0–3.32 and 1.0–3.7, respectively. For a long period range ($T = 0.8$–4.0 s), the horizontal acceleration response spectral value amplification coefficient of single-sided high and steep slopes decreases slightly and is in the range of 0.6–1.1, while the vertical acceleration response spectral value amplification coefficient stays at around 1.0. The horizontal and vertical acceleration response spectral value amplification coefficients of gentle and steep slope surfaces all stay at around 1.0.

To sum up, regardless of whether for single- or double-sided high and steep slopes, the horizontal and vertical acceleration response spectral values all amplify to a different extent, showing that the amplification effects of double-sided high and steep slopes on the acceleration response spectrum are greater than the amplification effects of single-sided high and steep slopes, the horizontal amplification effects of the acceleration response spectrum are greater than the vertical amplification effects, and the amplification effects of the acceleration response spectrum on steep slope surfaces are greater than those on gentle slope surfaces. Therefore, the slope topography and bevel angles have a significant influence on the changing rules of the slope surface acceleration response spectrum along elevations.

4.5 Local Topography Effect

To study seismic dynamic response rules of rock slopes, analysis from the perspective of the whole slope was conducted by seismic array monitoring, shaking table test and the above numerical analysis on individual independent slopes and continuous slopes of dozens of meters, hundreds of meters, or even much longer. The field survey results of slope hazards for the Wenchuan earthquake show that large amounts of local topography, such as independent mountains, weak mountain ridges and the turning parts of slope shape, have intense dynamic response and serious hazards under strong seismic effects, with obvious local topography effects. At the same time, the field monitoring results of slope seismic dynamic response also show that the seismic ground motion amplification effects of micro topography with multiple free surfaces is obvious. This indicates that local topography and micro topography perform important control functions in seismic dynamic response. Thus, this chapter will continue to focus on continuous slopes of dozens of meters, hundreds of meters, or longer, and will analyze the models in general and investigate the influence of slope tops with different forms on PGA elevation amplification effects so as to enrich research results of rock slope seismic dynamic characteristics from a micro perspective.

4.5.1 Key Problems in Numerical Analysis

This chapter will adopt slope models with different forms of slope tops (see Figs. 4.19, 4.20, 4.21 and 4.22). The slope height is about 80 m, with bevel angles of about 45°, and the slope tops of different forms have the same or similar elevation. At the same time, test points of horizontal acceleration, horizontal velocity and horizontal displacement are set up evenly on slope top position. The real seismic waves of the Wenchuan Wolong seismic array are selected simultaneously with horizontal and vertical PGA of 957 cm/s/s and 857 cm/s/s, and a duration of 163 s. See Figs. 4.23 and 4.24 for detailed horizontal and vertical acceleration time-history curves. In addition, the rock and soil materials select the materials of the prototype slope of the above single- and double-sided rock slope shaking table test. See Fig. 4.5 for concrete parameters (Table 4.5).

Fig. 4.19 Numerical analysis model of flattop slopes

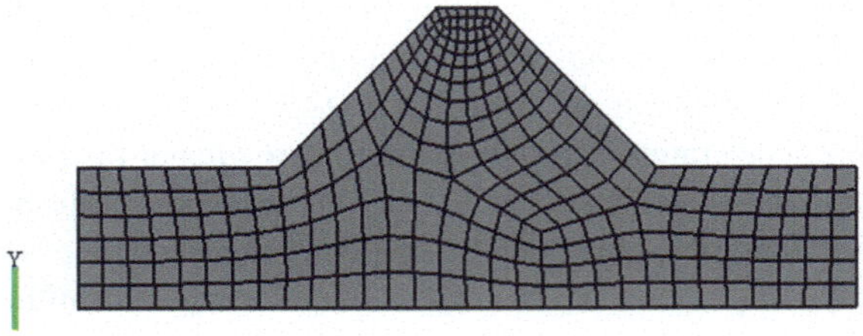

Fig. 4.20 Numerical analysis model of single-peak slopes

Fig. 4.21 Numerical analysis model of double-peak slopes

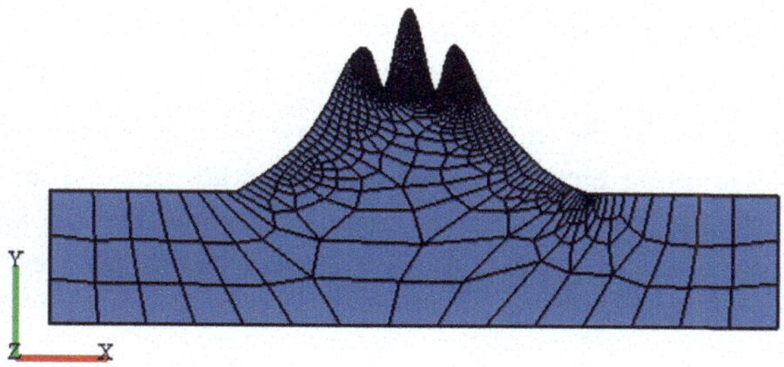

Fig. 4.22 Numerical analysis model of three-peak slopes

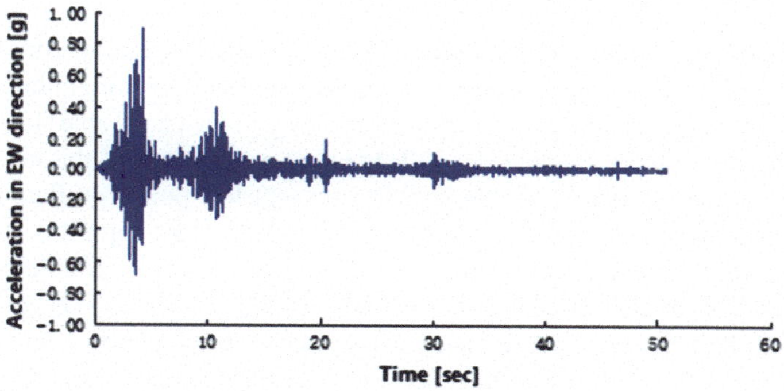

Fig. 4.23 Time-history curve of horizontal acceleration

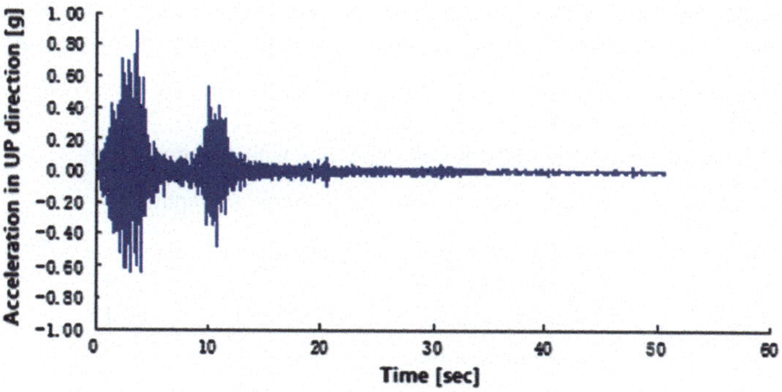

Fig. 4.24 Time-history curve of vertical acceleration

Table 4.5 Material parameters of rock and soil mass

Lithology	Elasticity modulus (E/MPa)	Poisson's ratio (μ)	Cohesive force (c/kPa)	Internal friction angle (°)	Unit weight (γ/kN/m^3)
Rock mass	450	0.25	0.936	37.9	22.0

4.5.2 Calculation Results Analysis

In order to further study the influence of slope surface forms on acceleration amplification effects along elevations, this chapter sorts out and analyzes the calculation results of test points 1, 2 and 3 on every model top and carries out investigations from the perspective of PGA, acceleration Fourier spectrum and acceleration response spectrum.

4.5.2.1 Effects of Slope Top Forms on Acceleration Peak Value Amplification Effect Along Elevations

To further study the influence of slope surface forms on acceleration amplification effects along elevations, this chapter picks up the aptitude of horizontal displacement, velocity and acceleration of every monitoring point (see Table 4.6).

Through comprehensive analysis of Table 4.6, we know that the acceleration peak values of all test points are above 2.0 g (except test points on flattop #2), in which the peak acceleration of test points on three-peak slope top #2 is 41.2 m/s^2. Compared with the peak acceleration of 9.57 m/s^2 of the input seismic wave, the peak acceleration of different slope forms amplify 2.0–4.5 times. The monitoring results of the flattop show that the peak acceleration of test points 1 and 3 on the two sides are 21.35, 18.65 and 21.35 m/s^2, which indicates that the seismic response near the free face direction is more intense. The acceleration amplitudes of the three-peak slope top, double-peak slope top, single-peak slope top and flattop are 41.2, 39.35, 25.1 and 21.35 m/s^2, respectively, in which the maximum peak acceleration of the three-peak slope top is about twice that of the flattop, fully explaining why the local topography of the slope top has a significant influence on

Table 4.6 Aptitude response characteristics of slope tops with different forms

Slope top topography	Horizontal aptitude	Slope top aptitude		
		1	2	3
Flattop	Displacement (m)	0.147	0.146	0.147
	Velocity (m/s)	0.739	0.717	0.739
	Acceleration (m/s^2)	21.35	18.65	21.35
Single-peak slope top	Displacement (m)	0.153	0.169	0.153
	Velocity (m/s)	0.818	0.970	0.818
	Acceleration (m/s^2)	20.4	25.1	20.4
Double-peak slope top	Displacement (m)	0.171	0.153	0.171
	Velocity (m/s)	1.61	1.42	1.61
	Acceleration (m/s^2)	39.35	27.4	39.35
Three-peak slope top	Displacement (m)	0.15	0.21	0.15
	Velocity (m/s)	1.47	1.74	1.43
	Acceleration (m/s^2)	30.2	41.2	30.95

the intensity of the seismic ground motion response. Through comprehensive analysis of the maximum peak velocity, acceleration and displacement, comparing results of test points on different slope top forms at the same elevation, we know that the seismic ground motion response rule is as follows: the three-peak slope top is greater than the double-peak slope top, which is greater than the single-peak slope top, which is greater than the flattop.

4.5.2.2 Influence of Slope Top Forms on Acceleration Fourier Spectrum Amplification Effect Along Elevations

To study the influence of slope top forms on the acceleration Fourier spectrum, based on the previous research results regarding acceleration amplitude, in this chapter we will select test points with the maximum acceleration amplitude, i.e. #1 of flattop, #2 of single-peak slope top, #1 of double-peak top and #2 of three-peak top. See Figs. 4.25, 4.26, 4.27 and 4.28 for detailed calculation results.

We know from comprehensive analysis of Figs. 4.25, 4.26, 4.27 and 4.28 that the slope top forms have a major influence on the frequency spectrum components of the slope top acceleration Fourier spectrum, mainly as follows: the high

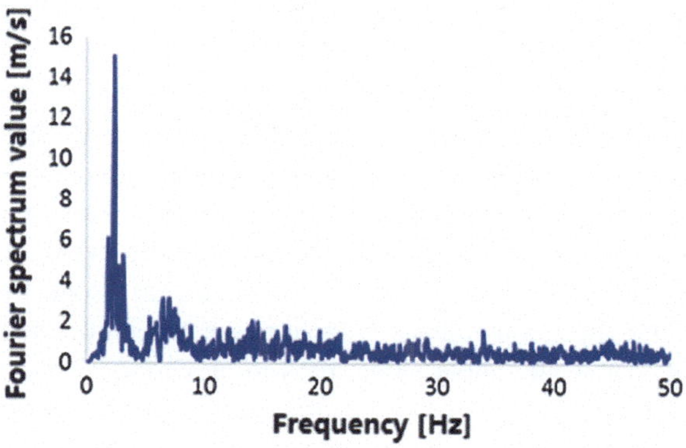

Fig. 4.25 Acceleration Fourier spectrum of test point #1 on flattop

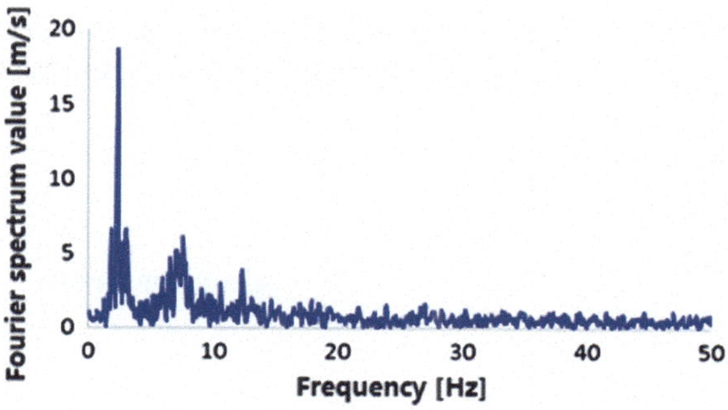

Fig. 4.26 Acceleration Fourier spectrum of test point #2 on single-peak slope top

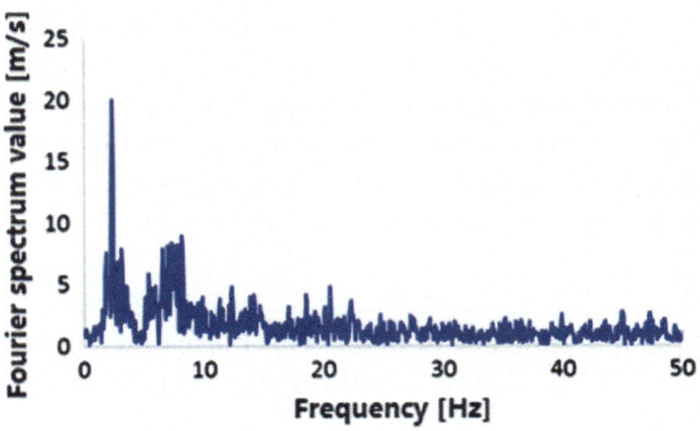

Fig. 4.27 Acceleration Fourier spectrum of test point #1 on double-peak slope top

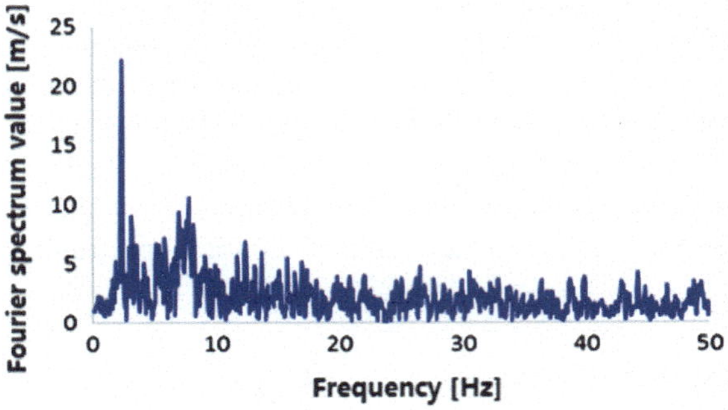

Fig. 4.28 Acceleration Fourier spectrum of test point #2 on three-peak slope top

frequency components of the acceleration Fourier spectrum on slope tops of flattop, single-, double- and three-peak slopes are gradually strengthened, and the second step frequency gradually becomes obvious. This may be caused by the horizontal section area of slopes with flattop changing evenly along elevations, while that of slopes with single-, double-, or three-peak slope tops suddenly decreases at the slope tops, which leads to large phase differences between slope top action and the action of the whole slope mass in an earthquake, even causing a whiplash effect, resulting in further large peak acceleration, more high frequency components and more obvious second step frequency.

4.5.2.3 Influence of Slope Top Forms on Acceleration Response Spectrum Amplification Effect Along Elevations

To study the influence of slope top forms on the acceleration response spectrum, in this chapter we will still select test point #1 of the flattop, #2 of the single-peak slope top, #1 of the double-peak top and #2 of the three-peak top. See Figs. 4.29, 4.30, 4.31 and 4.32 for detailed calculation results.

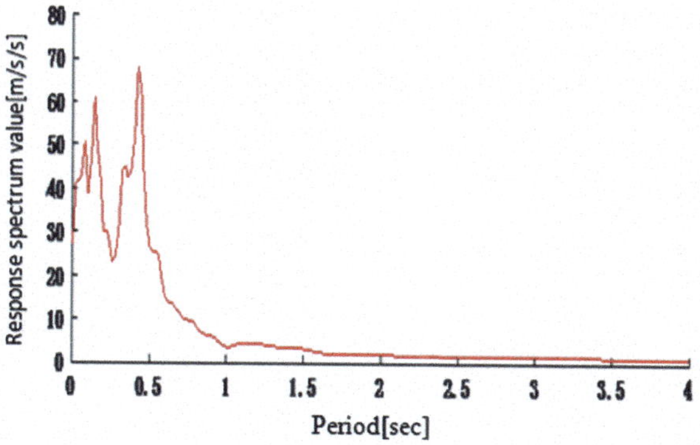

Fig. 4.29 Acceleration response spectrum of test point #1 on flattop

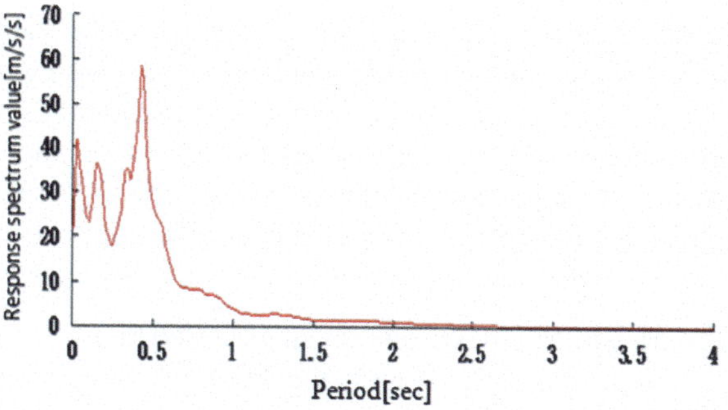

Fig. 4.30 Acceleration response spectrum of test point #2 on single-peak slope top

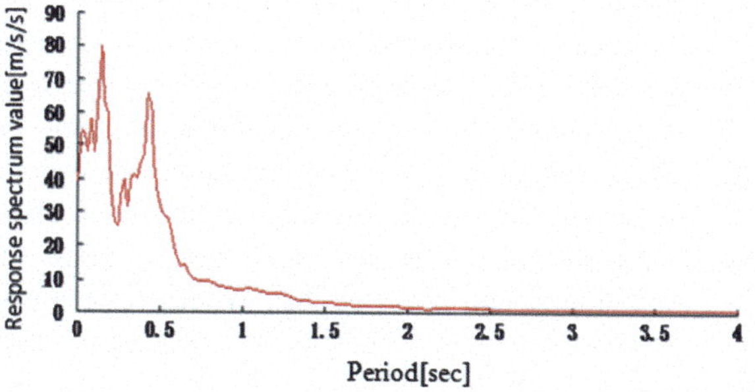

Fig. 4.31 Acceleration response spectrum of test point #1 on double-peak slope peak

We know from comprehensive analysis of Figs. 4.29, 4.30, 4.31 and 4.32 that the slope top forms have a major influence on the frequency spectrum components of the slope top acceleration response spectrum, which is mainly as follows: the characteristic periods of acceleration response spectrum of flattop, single-, double- and three-peak slopes are gradually transferred from long period to short period, and the double peak value of the response spectrum gradually become obvious, which is basically identical to the calculation results of the Fourier spectrum.

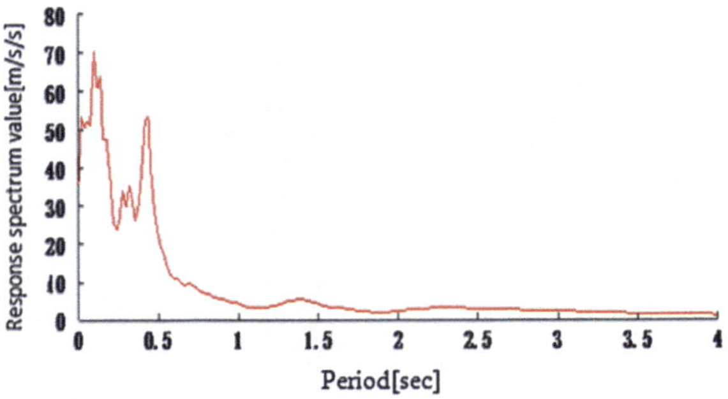

Fig. 4.32 Acceleration response spectrum of test point #2 on three-peak slope top

4.6 Brief Summary

In this chapter we have focused on the two major landslide hazard spots on the left side of G213, and have built a numerical analysis model based on the shaking table test model, and used the new numerical calculation software, GDEM, to investigate the general dynamic response characteristics of single- and double-sided high and steep rock slopes under intense seismic effects. At the same time, we have investigated the local topography effects of rock slope seismic dynamic response rules on a micro level, and have come to the following conclusions:

(1) Regardless of whether for single-sided high and steep slopes or for double-sided high and steep slopes, the PGA of different positions along elevations all amplify to a different extent, as follows: the amplification effects of double-sided high and steep slopes on acceleration are greater than those of single-sided high and steep slopes; PGA amplification effects on the slope surface are greater than those inside the slope mass; the horizontal acceleration coefficient is greater than the vertical acceleration coefficient; and the acceleration coefficient on steep slope topography is greater than that on gentle slope topography.
(2) The acceleration Fourier spectra along elevations of single- and double-sided high and steep slope surfaces all show the changing rules of low frequency component amplification, and the changing rules are influenced by slope topography and bevel angles. Whether for horizontal acceleration or vertical acceleration, the acceleration response spectrum along elevations of single-sided slope surfaces shows a single peak value, while that of the double-sided slope surfaces follows a process of converting from a single peak value to a double peak value.
(3) The local topography of the slope top has a significant influence on the intensity of the seismic response, and the detailed rules are as follows: the three-peak slope top is greater than the double-peak slope top, which is greater than the single-peak slope top, which is greater than the flattop. All of their high frequency components gradually strengthen, and the second step

frequency gradually becomes obvious. At the same time, the characteristic periods of acceleration response spectra of flattop, single-, double- and three-peak slopes are gradually transferred from long period to short period, and the double peak value of the response spectrum gradually becomes obvious. The peak accelerations of the different slope topographies all amplify 2.0–4.5 times, and the maximum peak acceleration of the three-peak slope top is about twice that of the flattop.

Chapter 5
Theoretical Solutions of Acceleration Amplification of Hill

For the seismic ground motion characteristics of rock slopes, comprehensive and in-depth research has been conducted through seismic array monitoring at Xishan Park, Zigong, shaking table test, and numerical analysis from the qualitative perspective, and the results show that the seismic ground motion response of double-sided rock slope is more intense than that of single-sided rock slope. In practical engineering, to evaluate something comprehensively and systematically needs qualitative cognition and quantitative analysis. Both of them complement each other, and none is dispensable. However, as for the quantitative analysis method for rock slope acceleration elevation amplification effects, some of the theoretical solutions came out in recent years can only consider the effect of simple harmonic waves, but not consider the time–frequency effects of seismic waves, i.e., the seismic wave frequency and amplitude change with time. While seismic wave is a nonlinear non-stationary signal, its frequency and amplitude gradually change over time. Therefore, this chapter, based on the previous qualitative research results, using Hilbert–Huang Transform (HHT for short) combining with elastic wave theory and double-sided rock slope analysis model, proposes the time–frequency analysis solution of rock slope acceleration elevation amplification effects that achieves transition from qualitative analysis to quantitative analysis of acceleration amplification effects so as to provide accurate reference to research and design of slope seismic engineering.

5.1 Formula Derivation

The double-sided rock slope acceleration elevation amplification effects time–frequency method is derived, based on the elastic wave theory and horizontal slice method, through building force equilibrium differential equation of micro-unit, using Hilbert–Huang Transform and boundary conditions, and applying modal analysis method and normal mode theory.

Y. Changwei et al., *Slope Earthquake Stability*,
DOI 10.1007/978-981-10-2380-4_5

5.1.1 General Thinking and Basic Assumption

The time–frequency analysis method derivation of double-sided rock slope acceleration elevation amplification effects mainly follows the following process: firstly, analyze the model and divide it into micro-units of N horizontal slices; then, select anyone of the micro-units; use HHT and elastic wave theory to solve the wave power when seismic wave passing through it; build force equilibrium differential equation and combine boundary conditions; and lastly apply modal analysis method and normal mode theory to work out time–frequency analysis solution of double-sided rock slope acceleration elevation amplification effects. The derivation of this method is based on the following hypothesis: ① The slope materials are homogeneous and isotropic elastomer; ② the bedrock is of rigid body so we can pay no attention to the bedrock deformation; ③ due to that, wave-type conversion will occur when seismic wave arrive at the slope surface with its major energy transferring into Rayleigh surface wave moving along the slope surface, and Rayleigh surface wave moves with elliptic polarization, so there will turn out two equal stress component of opposite direction within the range of one elliptic polarization. This method supposes that the slope material is homogeneous and isotropic elastomer. Therefore, the total stress component produced by Rayleigh wave on the slope surface is near 0, so the stress of Rayleigh wave will not be taken into consideration in formula calculation of this dissertation.

5.1.2 Generalized Analysis Model

Based on the hazard research results of Wenchuan earthquake, we generalize the analysis model of double-sided rock slopes (see Fig. 5.1). Figure 5.1 shows that at any elevation z, the shear displacement is v, shear stress is $\gamma = \partial v / \partial z$, the slope ratio of the two sides is 1:$n1$ and 1:$n2$, the total slope ratio is $n = n1 + n2$, the physico-mechanical parameters of slope mass are damping ratio c, density ρ, shear

Fig. 5.1 Generalized analysis model of double-sided rock slope

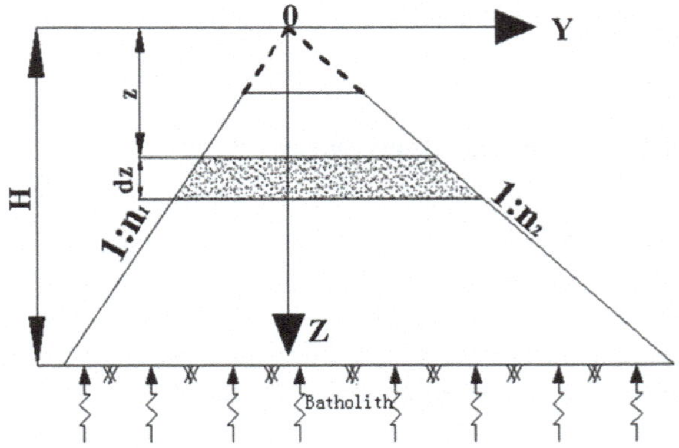

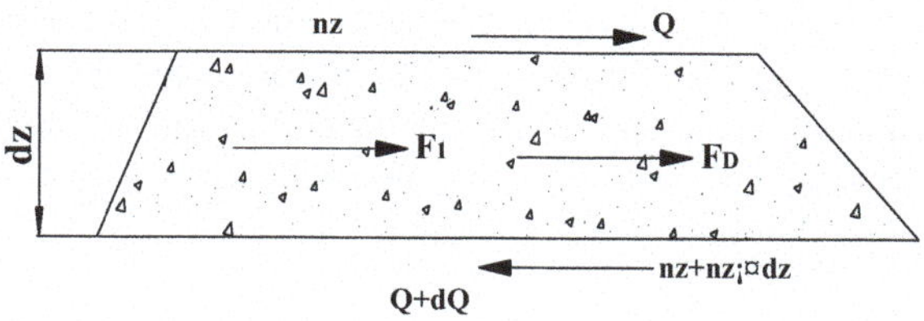

Fig. 5.2 Analysis model of micro-unit *A*

modulus G, and the acceleration time history of slope bottom input seismic wave is $\ddot{v}_g$. At the same time, divide the slope mass with horizontal slice method into N micro-units of horizontal slices and select micro-unit A as study object. Figure 5.2 is the force analysis model of micro-unit A under SV wave effects. The height of micro-unit A is dz, the width of the upper surface is nz, and the width of the lower surface is $nz + nz \cdot \mathrm{d}z$.

5.1.3 Derivation Process

Figure 5.2 shows that the top shear stress of micro-unit A is Q, the bottom shear stress is Q + dQ, the inertia force of earthquake on micro-unit A is $F1$, and the damping force is FD. See Formulas (5.1)–(5.4).

$$F_1 = \rho nz \cdot \mathrm{d}z\left(\frac{\partial^2 v}{\partial t^2} + \ddot{v}_g(t)\right); \tag{5.1}$$

$$F_D = cnz \cdot \mathrm{d}z \cdot \frac{\partial v}{\partial t} \tag{5.2}$$

$$Q = \tau nz = G\gamma nz = G\frac{\partial v}{\partial z}nz; \tag{5.3}$$

$$\frac{\partial Q}{\partial z}\mathrm{d}z = Gn\left(z\frac{\partial^2 v}{\partial z^2} + \frac{\partial v}{\partial z}\right)\mathrm{d}z \tag{5.4}$$

Build the force equilibrium equation of micro-unit A according to the stress conditions in Fig. 5.2 [see Formula (5.5)]. Then, bringing Formulas (5.1)–(5.4) into Formula (5.5), we get Formula (5.6):

$$F_1 + F_D - \frac{\partial Q}{\partial z}\mathrm{d}z = 0; \tag{5.5}$$

$$\frac{\partial^2 v}{\partial t^2} + \frac{c}{\rho}\frac{\partial v}{\partial t} - \frac{G}{\rho}\left(\frac{\partial^2 v}{\partial z^2} + \frac{1}{z}\frac{\partial v}{\partial z}\right) = -\ddot{v}_g(t) \tag{5.6}$$

Using separation of variables method is to make $v = \phi(z)Y(t)$, in which $\phi(z)$ is dimensionless vibration mode function, and $Y(t)$ is generalized coordinates or canonical coordinates, i.e., the time function of vibration mode. At the same time, separate Formula (5.6) into two differential equations only with variables t and z [see Formula (5.7)]:

$$\sum_{1}^{\infty}\phi\frac{\partial^2 Y}{\partial t^2}+\frac{c}{\rho}\sum_{1}^{\infty}\phi\frac{\partial Y}{\partial t}-\frac{G}{\rho}\left(\sum_{1}^{\infty}\frac{\partial^2\phi}{\partial z^2}Y+\frac{1}{z}\sum_{1}^{\infty}\frac{\partial\phi}{\partial z}Y\right)=-\ddot{v}_{\mathrm{g}}(t) \tag{5.7}$$

Multiplied by within $\sum$, we get Formula (5.8):

$$\sum_{1}^{\infty}\phi\rho z\phi\frac{\partial^2 Y}{\partial t^2}+c\sum_{1}^{\infty}\phi z\phi\frac{\partial Y}{\partial t}-\frac{G}{\rho}\left(\sum_{1}^{\infty}\rho z\phi\frac{\partial^2\varphi}{\partial z^2}Y+\frac{1}{z}\sum_{1}^{\infty}\rho z\phi\frac{\partial\varphi}{\partial z}Y\right)$$
$$=-\ddot{v}_{\mathrm{g}}(t)\rho z\phi \tag{5.8}$$

Because of mass $m = \rho nz \cdot \mathrm{d}z$, $c = cnz \cdot \mathrm{d}z$. Therefore, m and c are all proportional to z, so ρz can represent m and cz can represent c. According to orthogonality of vibration mode, we can know: When $m \neq n$, $\sum \varphi m m \varphi n = 0$, $\sum \varphi m c \varphi n = 0$; on the contrary, when $m = n$, $\sum \varphi m m \varphi n \neq 0$, $\sum \varphi m c \varphi n \neq 0$, so the above formula only suits for some type of vibration mode. In this way, we can separate $-\partial^2 Y/\partial t^2$, $\partial Y/\partial t$ and Y from $\sum$. $\ddot{v}_{\mathrm{g}}(t)$ has no relation to vibration mode, so they should not be in $\sum$ [see Formula (5.9)].

$$\left(\sum_{0}^{H}\varphi^2\rho z)\right)\frac{\partial^2 Y}{\partial t^2}+\left(\sum_{0}^{H}\varphi^2 cz\right)\frac{\partial Y}{\partial t}-\frac{GY}{\rho}\left(\sum_{0}^{H}\rho z\varphi\frac{\partial^2\varphi}{\partial z^2}+\frac{1}{z}\sum_{0}^{H}\rho z\varphi\frac{\partial\varphi}{\partial z}\right)$$
$$=-\left(\sum_{0}^{H}\rho z\varphi\right)\ddot{v}_{\mathrm{g}}(t) \tag{5.9}$$

Bring $c = 2\rho\lambda w$ into Formula (5.9) and then use each item in Formula (5.9) to divide $\sum_0^H \phi^2\rho z$ and $\frac{GY}{\rho}$, respectively, so Formula (5.9) can be separated into:

$$\frac{\partial^2 Y}{\partial t^2}+2\lambda w\frac{\partial Y}{\partial t}+R\frac{G}{\rho}Y=-\left(\sum_{0}^{H}\varphi z/\sum_{0}^{H}\varphi^2 z\right)\ddot{v}_{\mathrm{g}}(t)=-\eta\,\ddot{v}_{\mathrm{g}}(t) \tag{5.10}$$

$$\frac{\partial^2\phi}{\partial z^2}+\frac{1}{z}\frac{\partial\phi}{\partial z}+R\phi=0 \tag{5.11}$$

Due to $w^2 = \frac{k}{\rho}$, k of pure shear component is the function of G and component geometric size, so $\frac{RG}{\rho} = w^2$, in which R is the undetermined constant. So Formula (5.10) can be rewritten into Formula (5.12):

$$\frac{\partial^2 Y}{\partial t^2} + 2\lambda w \frac{\partial Y}{\partial t} + w^2 Y = -\left(\sum_0^H \varphi z / \sum_0^H \varphi^2 z \right) \ddot{v_g}(t) \tag{5.12}$$

5.1.3.1 Solution of Vibration Mode Function $\varphi(Z)$

Through comprehensive analysis of Formulas (5.11) and (5.12), it can be known that we need firstly to solve Formula (5.11) to get R before solving Formula (5.12). Now, we make $R = \left(\frac{\xi}{z}\right)^2$, so we will get Formulas (5.13) and (5.14):

$$\frac{\partial \phi}{\partial z} = \frac{\partial \phi}{\partial \xi} \cdot \frac{\partial \xi}{\partial z} = \frac{\xi}{z} \frac{\partial \phi}{\partial \zeta}; \tag{5.13}$$

$$\frac{\partial^2 \phi}{\partial z^2} = \frac{\xi^2}{z^2} \frac{\partial^2 \phi}{\partial \xi^2} \tag{5.14}$$

Bringing Formulas (5.13) and (5.14) into Formula (5.11), we will get Formula (5.15), while Formula (5.15) is zero-order Bessel function, whose general solution is Formula (5.16).

$$\frac{\partial^2 \phi}{\partial \xi^2} + \frac{1}{\xi} \cdot \frac{\partial \phi}{\partial \xi} + \phi = 0 \tag{5.15}$$

$$\phi = A_1 J_0(\xi) + A_2 N_0(\xi) = A_1 J_0(\sqrt{R}z) + A_2 N_0(\sqrt{R}z) \tag{5.16}$$

In Formula (5.16), J_0 is the first type of zero-order Bessel function and N_0 is the second type of zero-order Bessel function.

According to the boundary condition, z = 0, $\partial \phi / \partial z = 0$, take partial derivative of z from Formula (5.16), and we will get:

$$\partial \phi / \partial z = -\sqrt{R}[A_1 J_1(\sqrt{R}z) + A_2 N_1(\sqrt{R}z)] \tag{5.17}$$

In Formula (5.17), J_1 is the first type of zero-order Bessel function and N_1 is the second type of zero-order Bessel function.

Bring boundary conditions into Formula (5.17), and because when z approaches zero, $N_1(0)$ approaches infinity, and at the same time, when $z = 0$, $J_1(0) = 0$. Therefore, if we want to make the above formula equal to zero, then it must have $A_2 = 0$, so Formula (5.16) can be written into:

$$\phi = A_1 J_0(\sqrt{R}z) \tag{5.18}$$

Because vibration mode ϕ is the dimensionless relative value, we take $A_1 = 1$. According to the boundary condition $z = H, \phi = 0$, we bring it into Formula (5.18) and get $J_0(\sqrt{R}z) = 0$. Due to that, there are countless null points in the first type of zero-order Bessel function, so this dissertation adopts β_0, n to represent. Corner mark n represents modal order. At the same time, $\sqrt{R}H = \beta_{0,n}$, so $R = \beta_{0,n}^2/H^2$. Therefore, there are countless R values, and bringing them into Formula (5.18), we will get vibration mode function [see Formula (5.19)]. It is noteworthy that β_0, w appearing in the following texts all omit corner mark n.

$$\phi = A_1 J_0(\beta_0 \cdot \frac{z}{H}) \tag{5.19}$$

5.1.3.2 Solution of Vibration Mode Function of Time Y(t)

Through comprehensive analysis and simplification of Formula (5.12), we will get:

$$\frac{\partial^2 Y}{\partial t^2} + 2\lambda w \frac{\partial Y}{\partial t} + w^2 Y = -\left(\sum_0^H \varphi z / \sum_0^H \varphi^2 z\right) \ddot{v}_g(t) = -\eta \, \ddot{v}_g(t) \tag{5.20}$$

In Formula (5.20), $w^2 = \frac{RG}{\rho}, R = \frac{\beta_{0,n}^2}{H^2}$ and η are modal participation factors. For continuum, $\eta = \frac{\int_0^H \phi z \cdot \mathrm{d}z}{\int_0^H \phi^2 z \cdot \mathrm{d}z}$, and bringing into Formula (5.18), we can get modal participation factor:

$$\eta = \int_0^1 \frac{z}{H} J_0\left(\beta_0 \frac{z}{H}\right) d\left(\frac{z}{H}\right) / \int_0^1 \frac{z}{H} J_0^2\left(\beta_0 \frac{z}{H}\right) d\left(\frac{z}{H}\right) = \frac{2}{\beta_0 J_1(\beta_0)} \tag{5.21}$$

Because Formula (5.20) is a non-homogeneous, constant coefficient, second-order linear differential equation, its general solution is the general solution of its accompanied homogeneous equation plus the particular solution of an inhomogeneous equation [see Formula (5.22)].

$$Y = \frac{-\eta}{w'} \int_0^t \ddot{v}_g(\tau) e^{-\lambda w(t-\tau)} \sin w'(t-\tau) \mathrm{d}\tau \tag{5.22}$$

$$w' = \frac{\beta_0}{H} v_s \sqrt{1-\lambda^2} \tag{5.23}$$

5.1.3.3 Solution of Slope Mass Displacement, Velocity, and Acceleration Response

Seismic wave is a nonlinear non-stationary signal, and its frequency and amplitude gradually change over time. Therefore, based on the elastic wave theory, the seismic ground motion time history can be seen as the combination of N (the total number of sampling points) harmonic wave with different amplitude and frequency, so the harmonic equation of any time t (the mth sampling point) can be expressed by elastic displacement in an unified way [see Formula (5.24)], and the acceleration amplitude is given in Formula (5.25).

$$V_{\mathrm{g,m}}(t) = V_{\mathrm{g,m}}(0)\mathrm{e}^{i\omega_{\mathrm{m}}t}; \tag{5.24}$$

$$\left|\ddot{V}_{\mathrm{g,m}}(t)\right| = \left|\ddot{V}_{\mathrm{g,m}}(0)\omega_{\mathrm{m}}^2\right| = \ddot{V}_{\mathrm{g}}(\tau) \tag{5.25}$$

In Formulas (5.24)–(5.25), $V_{\mathrm{g,m}}(t)$ and $\left|\ddot{V}_{\mathrm{g,m}}(t)\right|$, respectively, represent the elastic displacement and acceleration amplitude of slope bottom seismic wave at t (the mth sampling point). $V_{\mathrm{g,m}}(0)$, $\left|\ddot{V}_{\mathrm{g,m}}(0)\right|$ and ω_{m}, respectively, represent the elastic displacement, acceleration amplitude, and instantaneous frequency slope bottom seismic wave at t (the mth sampling point).

Through comprehensive analysis of Formulas (5.19)–(5.26), we can get the time–frequency analysis solution of shear displacement response, velocity response, and acceleration response [see Formulas (5.26)–(5.28)].

$$v = \sum_{1}^{n} J_0\left(\beta_0 \frac{z}{H}\right) \cdot \frac{-\eta}{w'} \int_0^t \ddot{v}_{\mathrm{g}}(\tau)\omega_\tau^2 \mathrm{e}^{-\lambda w(t-\tau)} \sin w'(t-\tau)\mathrm{d}\tau \tag{5.26}$$

$$\frac{\partial v}{\partial t} = \sum_{1}^{n} J_0\left(\beta_0 \frac{z}{H}\right) \cdot \left[-\eta \int_0^t \ddot{v}_{\mathrm{g}}(\tau)\omega_\tau^2 \mathrm{e}^{-\lambda w(t-\tau)} \cos w'(t-\tau)\mathrm{d}\tau - \lambda\omega Y\right] \tag{5.27}$$

$$\frac{\partial^2 v}{\partial t^2} = -\ddot{v}_{\mathrm{g}}(t)\omega_\tau^2 + \sum_{1}^{n} J_0\left(\beta_0 \frac{z}{H}\right)\eta\Big[\frac{(1-2\lambda^2)w}{\sqrt{1-\lambda^2}} \int_0^t \ddot{v}_{\mathrm{g}}(\tau)\omega_\tau^2 \mathrm{e}^{-\lambda w(t-\tau)} \cdot \sin w'(t-\tau)\mathrm{d}\tau$$
$$+ 2\lambda w \int_0^t \ddot{v}_{\mathrm{g}}(\tau)\omega_\tau^2 \mathrm{e}^{-\lambda w(t-\tau)} \cdot \cos w'(t-\tau)\mathrm{d}\tau\Big] \tag{5.28}$$

5.1.4 Time–Frequency Analysis Process

Based on the above derivation, it can be known that the time–frequency effects of seismic waves mainly embody in the elastic displacement amplitude $\ddot{V}_{g,m}(t)$ and frequency ω_m of input seismic waves. Therefore, this dissertation the following analysis process of seismic wave time–frequency effects: firstly, break up seismic wave time–frequency into several of IMF (time history curve) through empirical mode decomposition (EMD for short); secondly, carry out channel switch, i.e., switch multi-channel signal composed of multiple IMF into single-channel signal made up of single IMF; thirdly, solve the instant frequency of each channel and draw the time–frequency curve of each IMF, and lastly, bring each IMF and their instant frequency into Formulas (5.26)–(5.28) to work out the results and superimpose the results, so we can get the seismic response of slopes such as displacement, velocity, and acceleration.

5.1.5 Practical Application

In order to more deeply and concretely explain the practical application of seismic wave time–frequency effects, this dissertation will, based on the analysis ideas of seismic wave time–frequency effects, conduct input wave time–frequency analysis, and take the example of NS direction component of Wenchuan wolong seismic wave to explain. The details are as follows: firstly, input the NS direction Wenchuan wolong seismic wave (see Fig. 5.3); secondly, carry out EEMD decomposition and get every IMF (see Fig. 5.4); and lastly, solve the instant frequency of every IMF (see Fig. 5.5). It is noteworthy that due to that, the duration of Wenchuan wolong seismic wave is a bit longer, so Figs. 5.3, 5.4, and 5.5 only

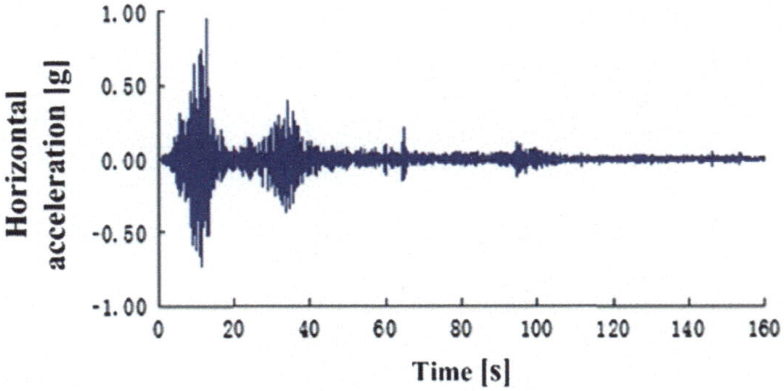

Fig. 5.3 NS direction Wenchuan wolong seismic wave

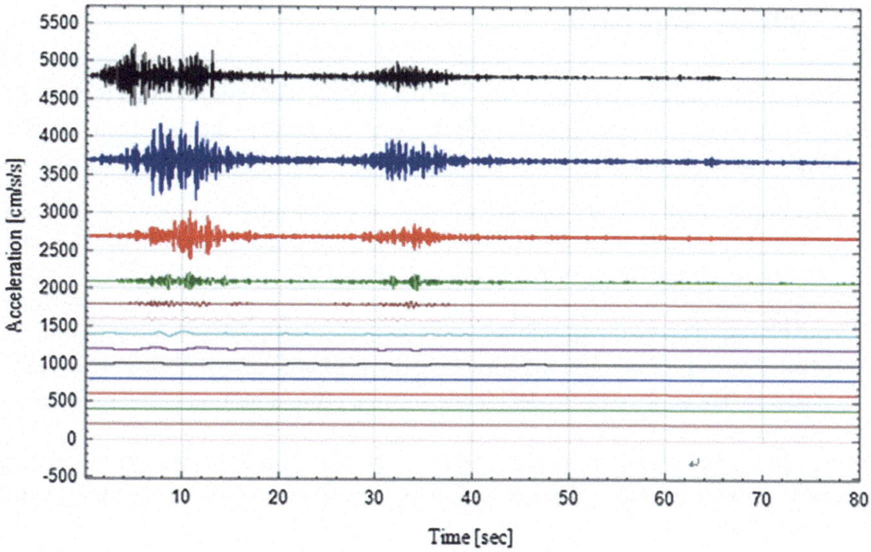

Fig. 5.4 IMF component of NS direction Wenchuan wolong seismic wave

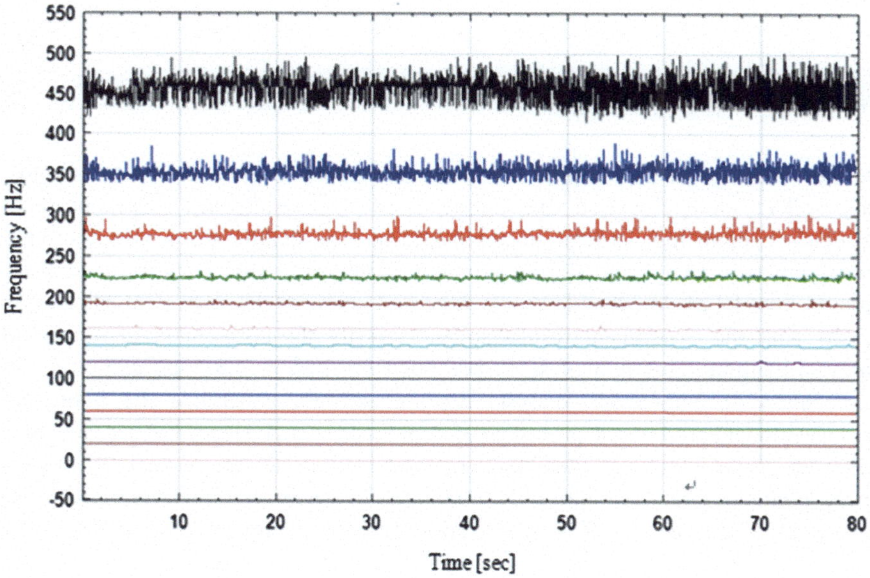

Fig. 5.5 IMF component instant frequency of NS direction Wenchuan wolong seismic wave

select calculation results of $T = 0$–80 s, including peak seismic ground motion acceleration, and demonstrate it clearly. At the same time, the Y coordinate in Fig. 5.5 only represents physical characters, but not size.

5.2 Solving Approach

To sum up, this chapter generalizes the solving approach of double-sided rock slope acceleration elevation amplification effects time–frequency analysis method, and the details are as follows:

(1) According to the study object, we sum up appropriate geological analysis model and decide the basic physical parameters such as geometric size and mechanical parameters of rock and soil mass;
(2) Decide the design response spectrum curve of this region according to Code for Seismic Design of Buildings, and synthesize artificial ground motion time history based on the seismic ground motion time history fitting method proposed by Hu et al., so as to decide the seismic ground motion acceleration time history of this region.
(3) Based on the seismic ground motion acceleration time history curve, work out IMF of bottom input seismic wave and the correspondent frequency time history curve;

Lastly, bring the obtained results into Formulas (5.26)–(5.28) to calculate and superimpose the results, so we can get the seismic response of slopes such as displacement, velocity, and acceleration.

5.3 Verification

Before verifying the correctness of double-sided rock slope acceleration amplification effects along elevation time–frequency analysis results, this chapter needs to explicitly explain the verification approach, and the details are as follows: first, build numerical analysis model by geometric similarity ratio of 1:1 to the shaking table test model of double-sided rock slope and compare the numerical calculation results with the experiment results so as to verify the correctness of numerical analysis method. This part of work has been finished in 4.3 Verification of the Correctness of Numerical Analysis Results; second, generalize the numerical analysis model based on the prototype of double-sided rock slope shaking table test, adopt the same numerical calculation method, and set up the acceleration monitoring points at slope top and mid-slope so as to verify the correctness of acceleration amplification effects along elevation time–frequency analysis method.

5.3.1 Overview

In order to accurately verify the correctness of acceleration amplification effects along elevation time–frequency analysis method, this dissertation will, based on the

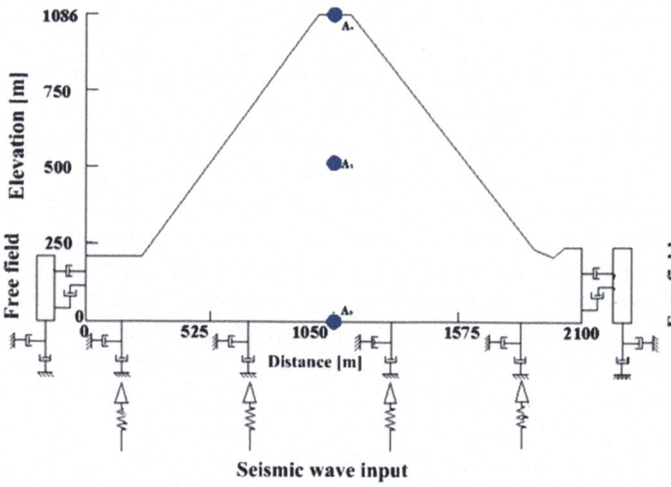

Fig. 5.6 Prototype slope model

Table 5.1 Physico-mechanical parameters of bedrock

Model name	Constitutive model	Unit weight	Cohesive force	Internal friction angle	Elasticity modulus	Poisson's ratio
Bedrock	Mohr-Coulomb	22 kN/m^3	93.6 kPa	37.9°	450 MPa	0.25

prototype of shaking table test model, restore the shaking table test model into the prototype slope according to the similarity system. It will pay no attention to the influence of slope volume, adjust the bevel angles on two sides into 50°, adopt the restored rock and soil mass parameters according to the similarity ratio, and input only the horizontal Wenchuan wolong seismic wave so as to produce only the vertical propagating shear wave. Set up horizontal acceleration and shear displacement monitoring points at the top and medium of prototype slope numerical model to observe the displacement and acceleration time history changes at top and medium positions under seismic effects. See Fig. 5.6 for numerical analysis model, Table 5.1 for physico-mechanical parameters of slope mass, and Figs. 4.23 and 4.24 for time history curve of Wenchuan wolong seismic elementary wave.

5.3.2 Verification

Based on the time–frequency analysis method solving approach of double-sided rock slope acceleration amplification effects along elevation (hereinafter referred to as time–frequency analysis method), bring the relevant parameters and the above calculation results into Formulas (5.26)–(5.28) to work out the shear displacement and horizontal acceleration of each point (see Tables 5.2 and 5.3 for detailed calculation results).

It can be known from Tables 5.2 and 5.3 that under the effects of Wenchuan wolong seismic wave with PGA = 0.2, 0.4, and 0.6, the numerical calculation results of shear displacement at slope top and mid-slope are all greater than calculation results worked out by time–frequency analysis method, and the error of these two distributed, respectively, in the range of 7.3–9.3 % and 4.8–8.6 %. The numerical calculation results of horizontal acceleration at slope top and mid-slope are all less than that of calculation results at slope top and mid-slope, and the error of these two distributed, respectively, in the range of 2.2–6.8 % and 2.0–4.1 %. Therefore, with the increase in height, the calculation error of time–frequency analysis method also increases; with the increase in PGA, the calculation error of time–frequency analysis method also increases, too. The reason for the above phenomenon may be that the time–frequency analysis method of double-sided high and steep rock slope acceleration elevation effects is derived based on the elastic wave theory without consideration of nonlinear characteristics of rock and soil mass. With the increase in seismic intensity, the nonlinear characteristics of rock and soil mass become more obvious, which weakens the acceleration amplification

Table 5.2 Comparison of time–frequency analysis method and numerical calculation of shear displacement and horizontal acceleration amplification coefficient at slope top

PGA	Shear displacement (m)			Horizontal acceleration amplification coefficient		
	Time–frequency analysis method	Numerical calculation	Error (%)	Time–frequency analysis method	Numerical calculation	Error (%)
0.2 g	0.51	0.55	7.3	2.12	2.073	−2.2
0.4 g	0.57	0.62	8.1	1.92	1.85	−3.7
0.6 g	0.64	0.706	9.3	1.73	1.62	−6.8

Table 5.3 Comparison of time–frequency analysis method and numerical calculation of shear displacement and horizontal acceleration amplification coefficient at mid-slope

PGA	Shear displacement (m)			Horizontal acceleration amplification coefficient		
	Time–frequency analysis method	Numerical calculation	Error (%)	Time–frequency analysis method	Numerical calculation	Error (%)
0.2 g	0.20	0.22	4.8	1.51	1.48	−2.0
0.4 g	0.26	0.28	7.1	1.35	1.32	−2.3
0.6 g	0.32	0.35	8.6	1.26	1.21	−4.1

Note Horizontal acceleration amplification coefficient in Tables 5.2 and 5.3 = acceleration of test point/horizontal acceleration of test point at model bottom

effects and strengthens shock insulation and shock absorption effects, which causes major difference in calculation results of these two.

In general, no matter the shear displacement or horizontal acceleration amplification coefficient, the calculation results of time–frequency analysis method, and the numerical analysis results of double-sided high and steep rock slope acceleration elevation effects are basically the same, with error controlled within 10 %, which fully verify the correctness of the calculation results and feasibility of calculation method in this dissertation.

5.4 Parameter Analysis

The previous research results based on field seismic array monitoring, shaking model test, and numerical analysis show that the amplitude and frequency of input wave and bevel angles have significant influence on rock slope acceleration amplification effects. Therefore, this chapter adopts time–frequency analysis method used by double-sided rock slope elevation amplification effects, and carry out research of influence of bevel angles, peak value of input wave and frequency on shear displacement and acceleration form the perspective of theoretical analysis.

5.4.1 Peak Value

In order to study the influence of input wave peak value on rock slope acceleration elevation amplification effects, now we take the following example to explain: The slope height is 1000 m, the bevel angles of the slope sides are all 60°, the damping ratio is 5 %, the elasticity modulus is 20000 Mpa, and the Poisson's ratio is 0.21. We input sine wave with PGA = 0.1, 0.2, 0.4, and 0.6, and its frequency is 0.8 Hz (see Figs. 5.7 and 5.8 for detailed calculation results). It is noteworthy that in Fig. 5.8, the acceleration amplification coefficient at slope top = slope top acceleration/slope bottom acceleration.

Through comprehensive analysis of Figs. 5.7 and 5.8, it can be known that with the effects of sine wave with PGA = 0.1, 0.2, 0.4, and 0.6, the acceleration amplification coefficient at slope top gradually decreases with the increase in input GPA, and it is within the range of 1.4–2.0. The maximum shear displacement gradually increases at the slope top, and it is within the range of 0.4–0.7 m.

5.4.2 Frequency

In order to study on the influence of input wave frequency on rock slope acceleration elevation amplification effects, now we adopt example of the same geometric

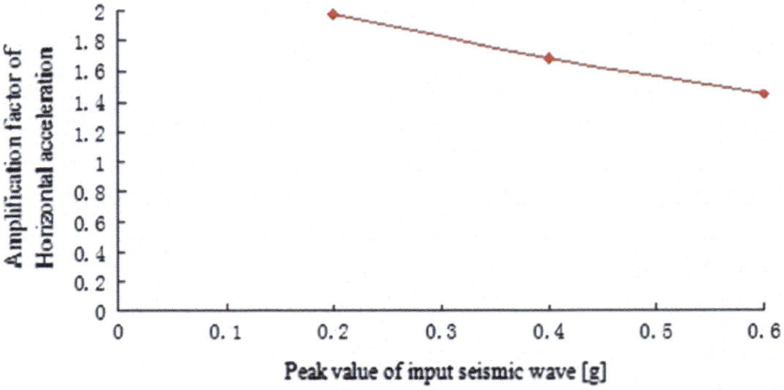

Fig. 5.7 Acceleration amplification coefficient at slope top versus input acceleration peak value

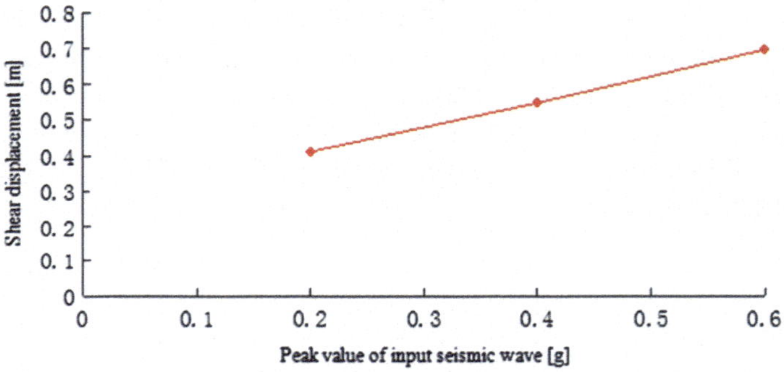

Fig. 5.8 Maximum shear displacement at slope top versus input acceleration peak value

size with the above example, and input sine wave with PGA = 0.4 g, whose frequency changes is as follows: 0.1, 0.5, 0.8, 1.0, 1.2, 1.4, 1.6, 1.8, 2.0, 2.2, 2.4, 2.7, 3.1, 3.4, 3.6, 4.0, 4.5, and 5.0 Hz (see Figs. 5.9 and 5.10 for detailed calculation results).

Through comprehensive analysis of Figs. 5.9 and 5.10, we can get the following rules: With the increase in the input wave frequency, the maximum shear displacement and acceleration amplification coefficient at the slope top all distribute in saddle shape and reach the maximum value with f = 1.6 Hz. This may be because when the input wave frequency is 1.6 Hz, it is close to the natural frequency (f = 1.56 Hz) of the slope mass, so resonance occurs in the slope mass and intensifies the seismic effects of the slope mass, which lead the maximum shear displacement and acceleration at the slope top reaching the maximum value at this input wave frequency.

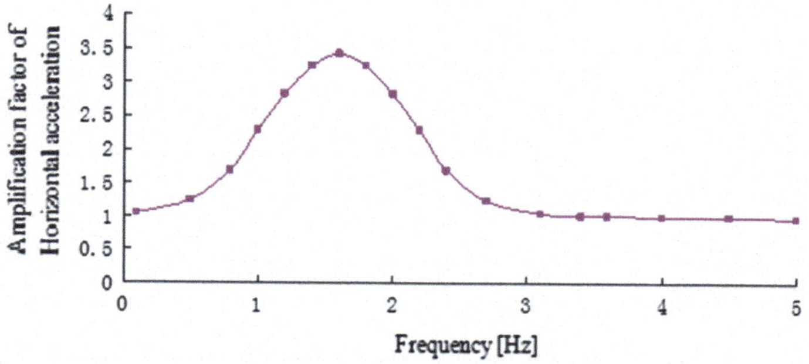

Fig. 5.9 Acceleration amplification coefficient at slope top versus input wave frequency

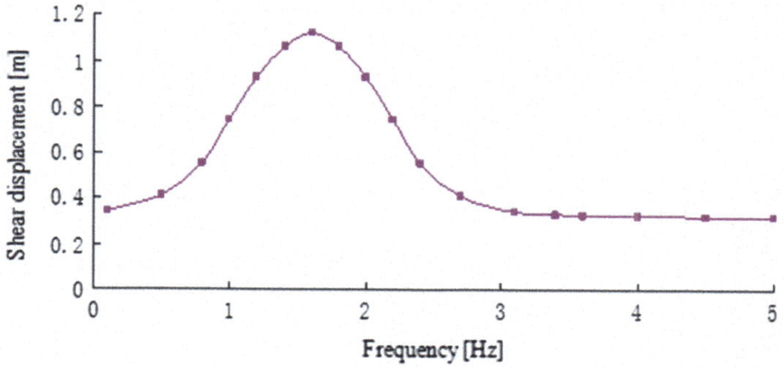

Fig. 5.10 Maximum shear displacement at slope top versus input wave frequency

5.4.3 Bevel Angles

In order to study the influence of bevel angles on rock slope acceleration elevation amplification effects, now we adopt example of the same geometric size with the above example, and input sine wave with PGA = 1.0 g and frequency = 1.5 Hz, and study on bevel angles of 30°, 35°, 40°, 45°, 50°, 55°, and 60° (see Figs. 5.11 and 5.12 for detailed calculation results).

Through comprehensive analysis of Figs. 5.11 and 5.12, we can get the following rules: With the increase in bevel angles, the maximum shear displacement and acceleration amplification coefficient at the slope top all gradually increase and distribute in "step-form"; i.e., in the ranges of 30°–45° and 50°–60°, the amplification coefficient all increases slightly, while in the range of 45°–50°, it increases suddenly, which is identical with the analysis results of the shaking table test and further explains the shaking table test results from the theoretical calculation perspective.

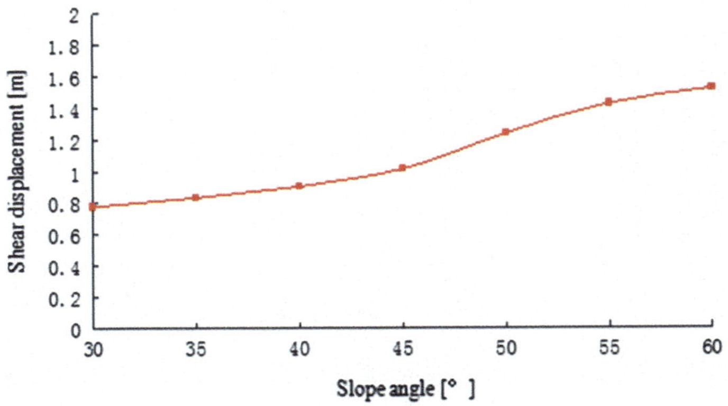

Fig. 5.11 Acceleration amplification coefficient at slope top versus input wave frequency

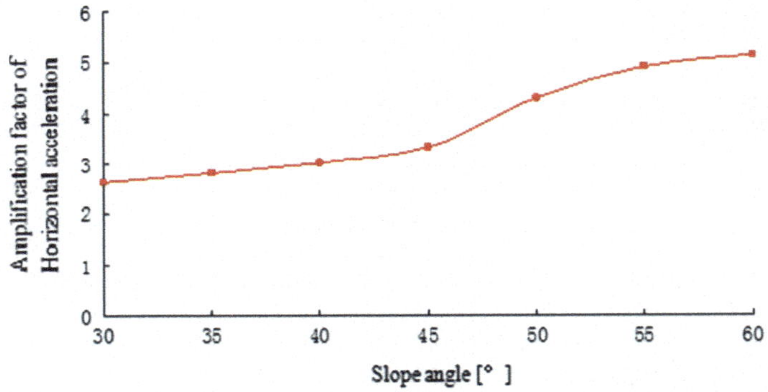

Fig. 5.12 Maximum shear displacement at slope top versus input wave frequency

5.4.4 "Whiplash Effect"

Under seismic effects, because the mass and stiffness of the protruding parts of the buildings are small, so there will come across faster velocity and larger displacement at moment of each seismic load turning back and forth, which is called "whiplash effect." While the top of high and steep rock slope would often come across the above phenomenon under strong seismic effects, and cause geometric hazards like sliding of the rock mass at the slope top, which has been verified in the field survey of geological disasters in Wenchuan earthquake. This chapter still selects the above example to study and exerts sine wave with PGA = 0.2, 0.4, and 0.6 g and with frequency = 1.6 Hz (see Fig. 5.13 for detailed calculation results). Through comprehensive analysis of Fig. 5.13, it can be known that "whiplash effect" occurs at the double-sided high and steep rock slope top when the input wave frequency is $f = 1.6$ Hz, and the shear displacement and acceleration

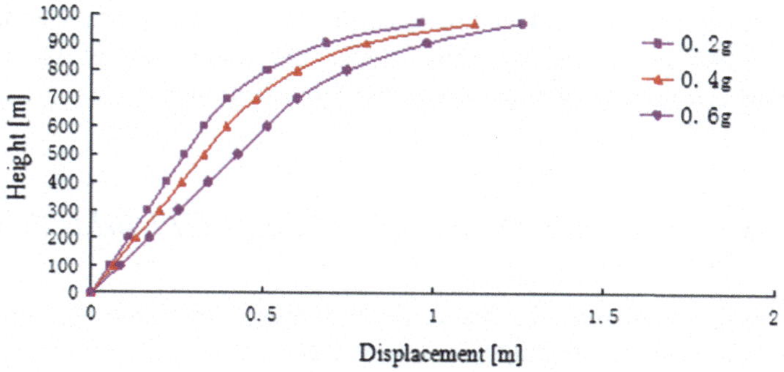

Fig. 5.13 Acceleration amplification coefficient at slope top versus input wave frequency

coefficient at the slope top all suddenly increase in multiple times. Therefore, the time–frequency analysis method of double-sided high and steep rock slope acceleration elevation amplification effects can explain the "whiplash effect" of slope mass from the perspective of theoretical calculation.

5.5 Comparative Analysis

The acceleration elevation amplification effect is the core part of slope engineering seismic design in the seismic code in China. This chapter, with full knowledge of rock slope seismic ground motion characteristics, focuses on the acceleration elevation amplification effects calculation method of some seismic design code in China and comparatively analyzes the differences between time–frequency analysis method of rock slope acceleration elevation amplification effects proposed in this chapter and the calculation methods of acceleration elevation amplification coefficient proposed in other codes, so as to put forward some modification suggestions on the calculation methods of acceleration elevation amplification effects in the seismic code in China.

5.5.1 Brief Introduction of Some Seismic Design Code

This chapter mainly studies the specifications and explanations about seismic amplification effects codes in China, which mainly include Code for Seismic Design of Buildings GB50011-2010 (Seismic Code for short), Specifications for Seismic Design of Hydraulic Structures DL/5073-2000 (Hydraulic Seismic Specifications for short) and Design Specification for Slope of Hydropower and Water Conservancy Project SL/386-2007 (Slope Design Specifications for short) [159–161]. It clearly states out in Seismic Code stressing the influence of topography on seismic amplification coefficient, while it puts forward in Hydraulic Seismic Specifications the amplification coefficient calculation of earth and rock dam. For the amplification coefficient of the 1st or 2nd hydraulic slopes, the regulations of Slope Design Specifications can refer to Hydraulic Seismic Specifications .

5.5.2 *Code for Seismic Design of Buildings GB50011-2010*

According to the regulation of Seismic Code, it should consider the amplification effects of earthquake on topographies such as slopes when build buildings, and the maximum acceleration amplification coefficient should multiply by certain amplifying coefficient to meet the design needs. The amplification coefficient value could multiply by amplifying coefficient in the range of 1.1–1.6 according to the concrete height and bevel angles of slopes. While the clause explanation makes it clear that the amplifying coefficient in the regulations is obtained through the summary of calculation results of seismic response analysis of large amount of rock and soil mass, the general situations can be reflected through seismic response analysis and large amount of slope seismic damage cases as follows:

1. The response is more intense as the height of high and protruding topography from the base level is higher;
2. The response is slighter as the distance between the slope and slope top boundary is larger;
3. From the perspective of rock and soil composition, the dynamic response of soil mass is stronger than that of the rock mass in the same terrain;
4. The response far away from the central part of the boundary becomes slighter as the top surface of high and protruding topography becomes broader;
5. The steeper the slope is, the stronger the amplification effects at its top is.

Based on the above qualitative changing trend, with slope elevation difference = H and tangent values of bevel angles = H/L, we put forward the calculation formula of slope acceleration amplification coefficient:

$$\lambda = 1 + \alpha \tag{5.29}$$

In this formula, λ is the acceleration amplification coefficient of the slope top;

α is the increase gradient of acceleration amplification coefficient of the slope top, and it is adopted in Table 5.4.

5.5.3 *Specifications for Seismic Design of Hydraulic Structures DL/5073-2000*

Clause 5.1.3 in Hydraulic Seismic Specifications specifically requires that the seismic stability calculation of earth and rock dam should adopt pseudo-static

Table 5.4 Seismic influence coefficient increase gradient of local protruding topography

Slope elevation difference $H(m)$	Non-rock topography	$H<5$	$5\leq H<15$	$15\leq H<25$	$H\geq 25$
	Rock topography	$H<20$	$20\leq H<40$	$40\leq H<60$	$H\geq 60$
Tangent values of bevel angles (H/L)	$H/L<0.3$	0	0.1	0.2	0.3
	$0.3\leq H/L<0.6$	0.1	0.2	0.3	0.4
	$0.6\leq H/L<1$	0.2	0.3	0.4	0.5
	$H/L\geq 1$	0.3	0.4	0.5	0.6

Note The maximum amplifying amplitude 0.6 regulated by the clause is given according to analysis results and comprehensive judgment, and this clause can be applied to all kinds of topographies including mounds, mountain ridges, cliffs, and steep slopes

method. The earth and rock dam are similar to slopes, so its dynamic amplification coefficient can refer to the seismic design of slopes. The clause regulates: When $H \leq 40$ m, the dynamic amplification coefficient distributes in trapezoid, with the maximum value occurring at dam top, and we take 1 for the dam bottom amplification coefficient and 2.0–3.0 for the dam top coefficient; when $H > 40$ m, the dynamic amplification coefficient $\beta(T) = 1 + [\beta(T)\max - 1]/3$ at 0–0.6*H* and distributes in trapezoid at 0.6*H*–1.0*H*, with the maximum value 2.0–3.0 occurring at dam top.

5.5.4 Design Specification for Slope of Hydropower and Water Conservancy Project SL/386-2007

Clause D.2.4 in Slope Design Specifications regulates that referring to the provisions in Hydraulic Seismic Specifications, the dynamic amplification coefficient of particles to slopes of 1st and 2nd grade could amplify from slope bottom to top after argument.

5.5.5 Comparative Analysis

Now, we take the example of numerical analysis model, rock and soil mass parameters, and external input ground motion in 5.3 Dual Side Rock Slope Acceleration Elevation Amplification Effects Time–frequency Analysis Method and use the regulated acceleration elevation amplification effects analysis method in Code for Seismic Design of Buildings GB50011-2010, Specifications for Seismic Design of Hydraulic Structures DL/5073-2000, and Design Specification for Slope of Hydropower and Water Conservancy Project SL/386-2007 to calculate the acceleration elevation amplification coefficient of this model (see Table 5.5 for

Table 5.5 Comparison table of calculation results of acceleration elevation amplification coefficient

		Calculation method		
Input PGA	Position	Method of this dissertation	Seismic code	Hydraulic Seismic Specifications
0.2 g	Mid-slope	1.51	1.6	1.33–1.67
0.4 g		1.35	1.6	Average value: 1.5
0.2 g	Slope top	2.12	1.6	2.0–3.0
0.4 g		1.92	1.6	Average value: 2.5

concrete results). Because the calculation results of Slope Design Specifications are identical with that of Hydraulic Seismic Specifications, this chapter will not do comparison of these two.

It can be known from Table 5.5 that under the effect of Wenchuan wolong seismic wave with PGA = 0.2 g, the acceleration elevation amplification coefficient at rock slope mid-slope is 1.51, which is slightly less than the calculation results of 1.6 in *Seismic Code,* and it is within the range of 1.33–2.0 in *Hydraulic Seismic Specifications*, which is close to average value of 1.5; the acceleration elevation amplification coefficient at slope top is 2.12, which is more than the calculation results of 1.6 in *Seismic Code*, and it is within the range of 1.33–2.0 in *Hydraulic Seismic Specifications* and less than its average value of 2.5. Under the effect of Wenchuan wolong seismic wave with PGA = 0.4 g, the acceleration elevation amplification coefficient at rock slope mid-slope is 1.35, which is slightly less than the calculation results of 1.6 in *Seismic Code,* and it is within the range of 1.33–2.0 in *Hydraulic Seismic Specifications* and less than its average value of 1.5; the acceleration elevation amplification coefficient at slope top is 1.92, which is more than the calculation results of 1.6 in *Seismic Code*, and it is within the range of 1.33–2.0 in *Hydraulic Seismic Specifications* and less than its average value of 2.5. It can be known from this that under the effects of seismic wave with PGA = 0.2 g, the calculation results of calculation method in this dissertation are basically identical with those in *Seismic Code*, and in *Hydraulic Seismic Specifications*, while under the effects of seismic wave with PGA = 0.4 g, the calculation results of calculation method for parts lower than mid-slope in this dissertation are less than those in *Seismic Code* and are identical with those in *Hydraulic Seismic Specifications;* and for parts higher than mid-slope, those in this dissertation is more than those in *Seismic Code* and less than those in *Hydraulic Seismic Specifications*. The reason for this can be that the methods in the specifications do not consider the changes of acceleration amplification with seismic intensity, and the importance of the industry served by each specification varies from each other. Therefore, it is suggested that the calculation of rock slope acceleration elevation amplification coefficient in *Seismic Code*, *Hydraulic Seismic Specifications* and *Slope Design Specifications should take into consideration the influence of seismic intensity, so as to improve the rationality of structural seismic design.*

5.6 Brief Summary

This chapter derives the time–frequency analysis method of double-sided rock slope acceleration amplification effects and verifies its rationality through shaking table test and numerical simulation. This method could not only take good consideration of the influence of three essential elements (PGA, frequency, and duration) on acceleration elevation amplification, but also explain the "whiplash effects" of slopes in the perspective of theoretical calculation, and this method comes to the following conclusions:

The double-sided rock slope acceleration elevation amplification effects time–frequency method is derived, based on the elastic wave theory and horizontal slice method, through building force equilibrium differential equation of micro-unit, using Hilbert–Huang Transform and boundary conditions, and applying modal analysis method and normal mode theory. The rationality of this method is verified through shaking model test and numerical analysis.

This method could not only take good consideration of the influence of three essential elements (PGA, frequency, and duration) on acceleration elevation amplification, but also explain the "whiplash effects" of slopes in the perspective of theoretical calculation.

Through carrying out research on the influence of bevel angles, input wave PGA, and frequency on shear displacement and acceleration, we come to the following rules: With the increase in seismic intensity, the shear displacement at slope top gradually increases and the acceleration amplification coefficient gradually decreases; the shear displacement of slope mass is in nonlinear distribution along elevation and suddenly increases in multiple times; with the increase in input wave frequency, the maximum shear displacement and acceleration amplification coefficient at slope top distributed in saddle shape, and reaching the maximum value when input frequency value equals the maximum value of the natural frequency of slope mass; with the increase in bevel angles, the maximum shear displacement and acceleration amplification coefficient all gradually increase and distribute in "step-form"; i.e., in the ranges of 30°–45° and 50°–60°, the amplification coefficient all increases slightly, while in the range of 45°–50°, it increases suddenly.

Through comparative analysis of high and steep rock slope time–frequency analysis method with the calculation results of acceleration elevation amplification effects analysis method regulated in *Code for Seismic Design of Buildings* GB50011-2010, *Specifications for Seismic Design of Hydraulic Structures* DL/5073-2000, and *Design Specification for Slope of Hydropower and Water Conservancy Project* SL/386-2007, it is suggested that the calculation of rock slope acceleration elevation amplification coefficient in *Seismic Code*, *Hydraulic Seismic Specifications,* and *Slope Design Specifications* should take into consideration the influence of seismic intensity, so as to improve the rationality of structural seismic design.

Chapter 6
Slope Deformation Characteristics and Formation Mechanism

The seismic stability of slopes has been the hot spot issues of geological disaster study, and especially after the 5.12 Wenchuan earthquake, it has become the study focus of specialists and scholars. While seismic instability mechanism is the key for slope seismic stability study, and the seismic ground motion characteristics is the precondition for seismic instability mechanism study. Therefore, based on the research results of seismic ground motion characteristics of slopes, it is quite necessary to conduct research on slope deformation characteristics and instability mechanism under seismic effects, and it has significant importance for post-disaster reconstruction.

After 5.12 Wenchuan earthquake, members in our research group immediately carried out large amount of research on collapses and sliding disasters in the seismic area and conducted statistical analysis to the total distribution of collapses and sliding disasters and distribution rules of classic regional collapses and sliding in the seismic area. Large amount of field survey results show that slope structure is still the main factor to control slope deformation and damage. Because of Wenchuan earthquake region located in deep canyon area with precipitous terrain, with effects of long period of gravity and rainfall, the middle and upper slope always develops thin cover layer and strong-weathered and relief relaxed rock mass, which is often located in the middle and upper slopes, and easily comes across instability damage under seismic effects. That makes the bedrock and overburden slopes made of soil layer and intensive weathered layer the most common slope structure with maximum instability quantity in seismic areas (see Figs. 6.1 and 6.2). Therefore, this chapter will, based on the previous research results of rock slope seismic ground motion characteristics, focus on bedrock and overburden slopes, select classic working condition, adopt approach of shaking table test combining with numerical analysis, and conduct in-depth study of bedrock and overburden slope deformation characteristics and formation mechanism under seismic effects from the perspective of time history and time–frequency.

Y. Changwei et al., *Slope Earthquake Stability*,
DOI 10.1007/978-981-10-2380-4_6

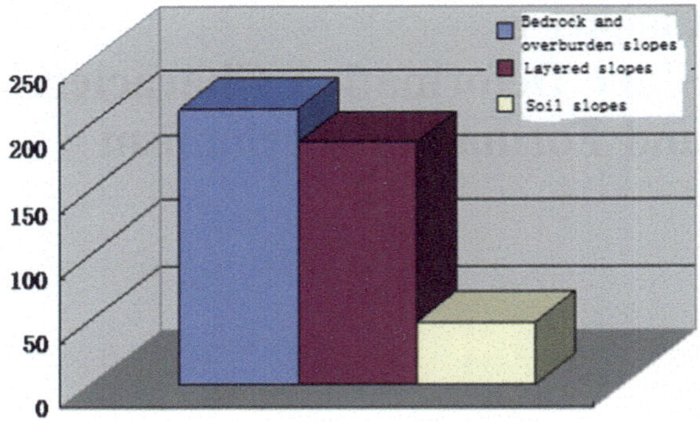

Fig. 6.1 Quantity distribution map of rock slope instability structure

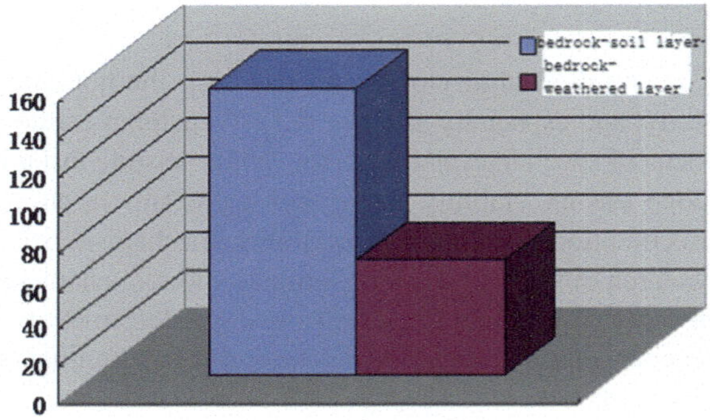

Fig. 6.2 Quantity distribution map of bedrock and overburden slope instability structure

6.1 Brief Introduction

Before conducting the research work, this chapter needs to explicitly explain the study approach, and the details are as follows: First, build numerical analysis model by geometric similarity ratio of 1:1 to the shaking table test model of double-sided rock slope and compare the numerical calculation results with the experiment results so as to verify the correctness of numerical analysis method. This part of work has been finished in Sect. 4.3. Second, for the prototype of double-sided rock slope shaking model test, adjust it appropriately according to the research focus in this chapter, generalize the numerical analysis model, adopt the same numerical calculation method with that of the prototype, and set up evenly large amount of test points of acceleration, displacement, stress and strain on sliding mass, sliding beds and the sliding mass surface, so as to conduct in-depth research on the deformation characteristics and formation mechanism of bedrock and overburden slope.

6.1.1 Numerical Simulation Method

In recent years, GDEM, a new numerical calculation software which can be well applied to geological disaster damage analysis, has developed. The software makes finite element method and discrete element method coupling together, i.e., using FEM inside the slope and DEM at the boundary of the slope, which can well solve the limits of FEM, finite difference method, and traditional DEM software. Therefore, this chapter adopts GDEM to carry out the simulation of progressive failure and movement process after instability of slope jointed rock mass under seismic effects.

6.1.2 Numerical Analysis Model

In order to deeply study on the seismic landslide mechanism of bedrock and overburden slope, this chapter builds numerical analysis model according to the prototype slope of shaking table test model, whose geometric characteristics are as follows: The height of single-sided bedrock and overburden slope is 1312 m, its width is 2240 m, bevel angle is 50°–60°, the width of the river valley at the slope foot is 103.8 m, the riverbed is slightly inclined, the bevel angle of the river valley right bank slope is in the range of 30°–40°, and the slope top is partly covered with seriously weathered layer; the height of the double-sided bedrock and overburden slope is 1086 m, and its width is 2100 m, which include steep slope topography and gentle slope topography with bevel angles, respectively, in the range of 50°–60° and 42°–50°. At the same time, the field survey results of slope sliding show that deposit body at mid-slope is the most serious, so we simplify the soil layer and weathered layer on the slope surface into deposit body at mid-slope according to the field survey results to better study the sliding mechanism of the slope. In addition, large amounts of monitoring points were set up inside and on the surface of the numerical model to monitor the dynamic response time history changes of slope acceleration, stress, and strain of sliding body and sliding mass under seismic effects. See Figs. 6.3 and 6.4 for details.

6.1.3 Numerical Constitutive Model and Material Parameters

In the calculation of DEM, the rock and soil mass of sliding bed and sliding mass mainly comes across plastic deform before instability, so elastoplastic model and Mohr–Coulomb failure principle (M-C failure principle) are adopted by both of them. After instability occurs, this chapter adopts rigid constitutive model for soil particles inside the sliding mass mainly because first, after instability occurs, this

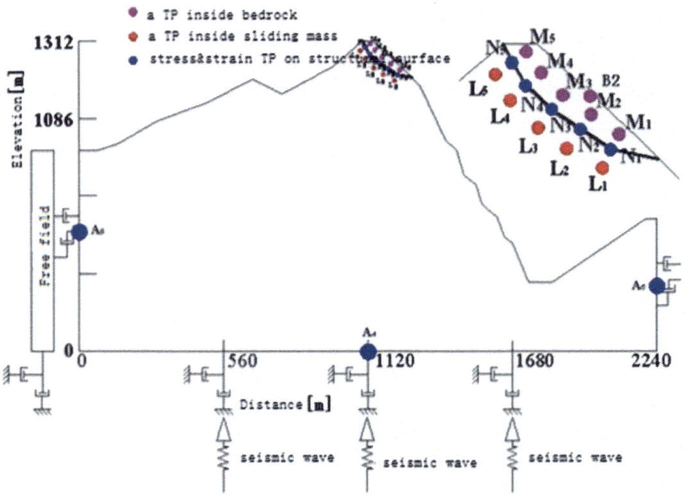

Fig. 6.3 The numerical analysis model of single-sided bedrock and overburden slope

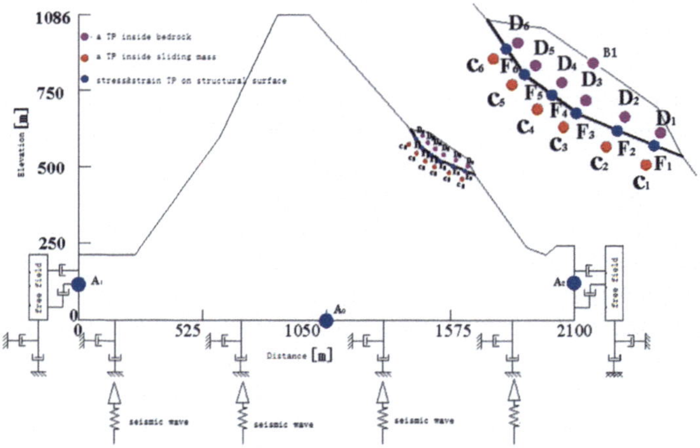

Fig. 6.4 The numerical analysis model of double-sided bedrock and overburden slope

chapter mainly focuses on the motion features of sliding mass such as the trajectory of sliding mass and deposition; second, because the displacement of sliding mass may amount to several hundred meters and deformation of the sliding mass differs many orders of magnitude to that, the sliding mass displacement is much more than deformation of soil particles inside the sliding mass; and third, if we still adopt Mohr–Coulomb failure principle after sliding occurs, it will cause slower convergence rate of slope system, which may lead to energy dissipation and non-convergence of system calculation. For sliding bed, because of its large geometric size and plastic deformation under seismic effects, this chapter, therefore, adopts elastoplastic model and M-C failure principle. Because failure phenomenon such as slips and cracks may occur on sliding mass structural surface and among soil particles inside sliding mass under seismic effects, this chapter will adopt brittle fracture constitutive model and M-C failure principle for all the mentioned positions. Adopt the stochastic structural surface model in GDEM to achieve order, complexity, and randomness of structural surface inside the sliding mass. Among blocks and blocks, it sets up joint-point mixed units to achieve independent calculation of particle motion of both structural surface spring and particle DEM, which can effectively solve the calculation of DEM initial states. The numerical simulation of this chapter adopts the exactly same material with the prototype slope, and the related parameters of sliding mass structural surface can be solved by relevant formulas in GDEM. See Table 6.1 for concrete physico-mechanical parameters and constitutive model of materials.

6.1.4 Boundary Conditions

The selection of boundary conditions in this part uses the same method with Sect. 4.2.3. In addition, because the Wenchuan wolong seismic station nears the prototype slope, this chapter selects the observed seismic wave by Wenchuan wolong seismic station with horizontal and vertical PGA of 957 cm/s/s and 857 cm/s/s and the duration of 163 s. See Figs. 4.23 and 4.24 for detailed horizontal and vertical acceleration time history curve.

6.2 Deformation Characteristics and Disaster Mechanism

In order to study the seismic sliding mechanism of single- and double-sided bedrock and overburden slopes, this chapter sets up acceleration Test Points L1–L5 and C1–C6 inside the slope mass of single- and double-sided prototype numerical analysis model, acceleration Test Points M1–M5 and D1–D6 inside the slope mass, stress and strain Test Points N1–N5 and F1–F6 on slope mass structural surface and displacement Test points B1 and B2, so as to conduct targeted study on dynamic characteristics and sliding mechanism of slope deposit body on single- and double-sided prototype slope under seismic effects.

6.2.1 Whole Landslide Process

In order to study the whole landslide process of bedrock and overburden slopes in earthquake, this chapter only sets up displacement monitoring points B1 and B2 on deposit body surface of single- and double-sided Fig. 6.5 for detailed results

Table 6.1 The physico-mechanical parameters of sliding mass, bedrock, and structural surface

Numerical analysis method	Constitutive model	Model name	Unit weight (kN/m^3)	Cohesive force (kPa)	Internal friction angle (°)	Elasticity modulus (MPa)	Poisson's ratio
Single- and double-sided bedrock and overburden slopes	M-C principle	Bedrock	22	93.6	37.9	450	0.25
		Slope deposit body	21	49.8	28.8	60	0.35
	Constitutive model	Structural surface location	Normal stiffness (MPa)	Shear stiffness (MPa)	Internal friction angle (°)	Tensile strength (kPa)	Cohesive force (kPa)
Single-sided bedrock and overburden slope	Brittle fracture constitutive model	Slope deposit body inside	3600	1800	32	25.98	48.8
		Structural surface	27,000	13,800	32	6.30	13.8
Double-sided bedrock and overburden slope		Slope deposit body inside	600	300	32	4.33	8.1
		Structural surface (slope foot)	4500	2300	32	1.21	2.3
		Structural surface (mid-slope)	4500	2300	32	1.05	2.3
		Structural surface (slope top)	6200	6200	36	3.86	7.2

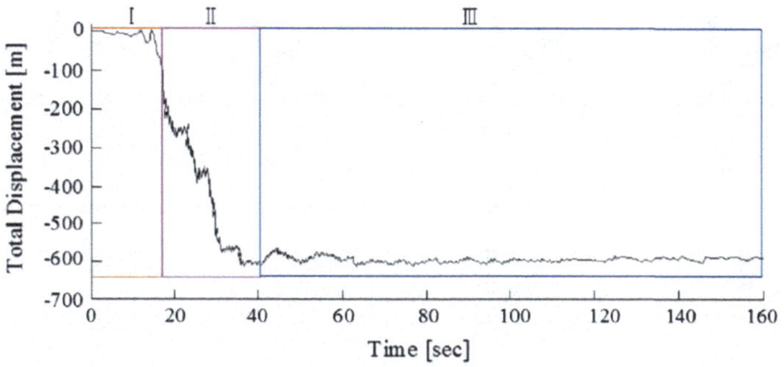

Fig. 6.5 The total displacement time history curve of B1

bedrock and overburden slopes. Due to the fact that the sliding process revealed by the monitoring results of B1 and B2 is basically the same, this chapter only selects the total displacement time history curve of B1 to analyze.

Figure 6.5 shows that the total displacement of B1 stays at around 0 when $T = 0\text{–}15.75$ s, while it suddenly increases at $T = 15.75\text{–}33.12$ s and its volatility gradually decreases at $T = 33.12\text{–}58.18$ s, ending at about 600 m. According to the above failure phenomenon, this chapter divides the sliding process into four phases: starting phase → high-speed sliding phase → deposit phase → self-stabilizing phase. Therefore, this chapter will select seismic response of starting phase at $T = 0\text{–}15.75$ s to further study the sliding mechanism.

6.2.2 Landslide Hazard Phenomenon

When GDEM carries out numerical solutions, it could update calculation mode in real time according to unit deformation and motion. Finally, shearing strain and tensile strain distribution of every structural surface inside the single- and double-sided bedrock and overburden slope models of different times in the earthquake was obtained through calculation (see Fig. 6.6). It is noteworthy that in Fig. 6.6, blue represents non-damage of contact elements, red represents shear failure, and green represents tensile failure.

Through comprehensive analysis of Fig. 6.6, it can be known that the whole process of deposit body sliding from deformation to failure of single- and double-sided bedrock and overburden slopes is basically the same, with a slight difference. Firstly, the difference of the two: At $T = 15.75$ s, the slope top sliding structural surface of single-sided bedrock and overburden slope is basically in complete connection, while the sliding mass at mid-slope of double-sided bedrock and overburden slope has already come across landslides, which fully demonstrates that with the same external conditions such as input seismic wave and sliding structural surface parameters, the seismic response of double-sided bedrock and

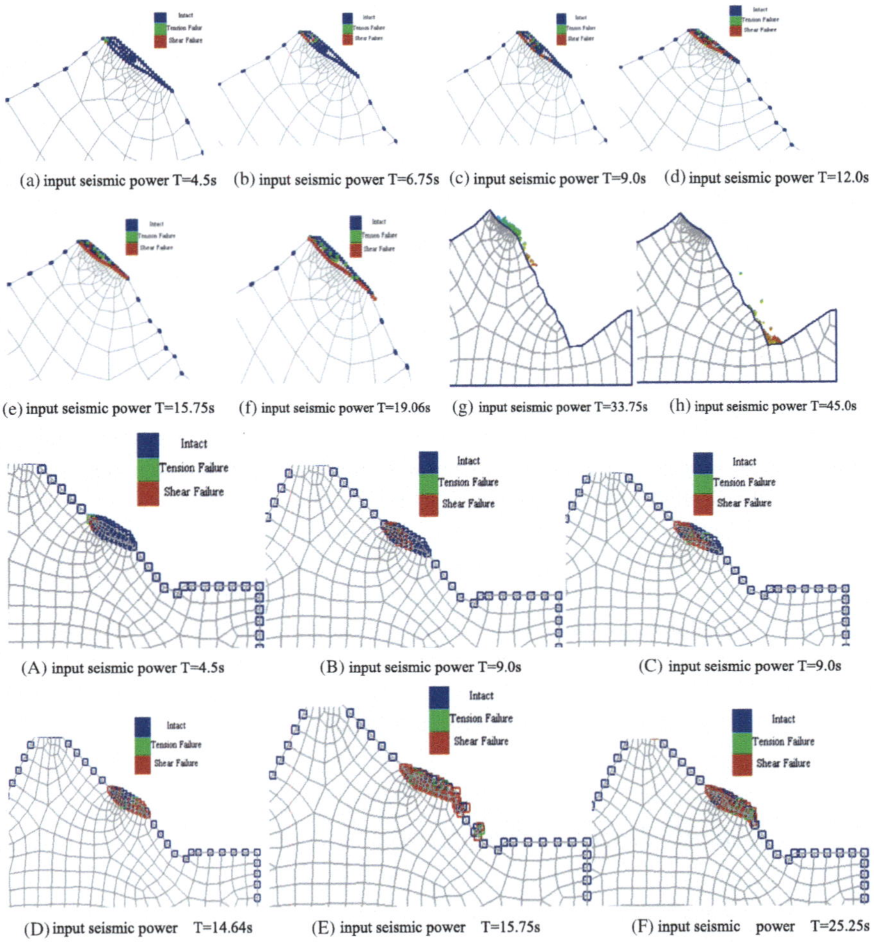

Fig. 6.6 Shear and tensile failure points' distribution map of single- and double-sided bedrock and overburden slopes. **a** Input seismic power $T = 4.5$ s, **b** input seismic power $T = 6.75$ s, **c** input seismic power $T = 9.0$ s, **d** input seismic power $T = 12.0$ s, **e** input seismic power $T = 15.75$ s, **f** input seismic power $T = 19.06$ s, **g** input seismic power $T = 33.75$ s, **h** input seismic power $T = 45.0$ s, **A** input seismic power $T = 4.5$ s, **B** input seismic power $T = 9.0$ s, **C** input seismic power $T = 9.0$ s, **D** input seismic power $T = 14.64$ s, **E** input seismic power $T = 15.75$ s, and **F** input seismic power $T = 25.25$ s

overburden slopes is stronger than that of the single-sided bedrock and overburden slopes, and the seismic damage is also more serious. The above phenomenon is identical not only with the research results of rock slope seismic ground motion characteristics, but also with the survey results of Wenchuan earthquake damage. Secondly, the similarity of the two: First, at the beginning of seismic ground motion, a few tensile and shear damage failure points occur at the sliding mass top; with the increase in input ground motion, the shear failure points at the sliding mass structural surface gradually develop downward and form large amount of discontinuous tensile and shear failure points; with the continuous increase in input ground motion, the tensile and shear failure points on the sliding mass structural surface gradually connect and form continuous shear failure slip crack surface and develop to claviculate section of sliding mass leading edge; lastly, with input seismic wave acceleration reaching the peak value, the shear failure points on the sliding mass structural surface rapidly extend and form continuous shear crack surface connecting the whole sliding mass structural surface. The above failure phenomenon is identical not only with the shaking table test phenomenon, but also with the research results of most sliding mass occurrence time in 5.12 Wenchuan earthquake. The reasons for the above phenomenon may be as follows: (1) At the beginning of gravity and seismic power, tensile stress concentration occurs at the sliding mass top, causing tensile deformation of sliding mass along sliding mass structural surface trailing edge, which causes tensile failure at this position; (2) under seismic effects, shear strength inside sliding bed and on the sliding mass surface will gradually decrease with the increase in seismic ground motion acceleration (see Sect. 6.2.3.3 for details); (3) with the increase in seismic ground motion acceleration, the inconsistency of motion between sliding bed and sliding mass gradually increases which intensifies the development of shear and tensile failure points on sliding mass structural surface (see Sect. 6.2.3.4); and (4) with the increase in seismic ground motion acceleration, the difference of seismic energy distribution and dissipation gradually enlarges between sliding bed and sliding mass, which intensifies the damage of sliding mass structural surface and causes slope sliding. See Sect. 6.2.3.5.

To sum up, the bedrock and overburden slopes coming across landslides is a balanced and progressive process. Under the effects of gravity and seismic power, tensile stress concentration firstly occurs at sliding mass top, causing deformation of sliding mass along sliding mass structural surface trailing edge, which causes tensile and shear failure points at this position. Then, with continuous seismic power, the shear failure points on sliding mass structural surface gradually develop to claviculate section of sliding mass leading edge with increasing amount of tensile failure points on sliding mass surface, which finally causes progressive damage of claviculate section, crack surface connecting to sliding zone, and sliding mass slipping from shear crack into landslide.

6.2.3 Landslide Mechanism Analysis

This chapter will, based on the previous numerical calculation results, analyze from the micro-perspective of stress–strain development on sliding mass structural surface in earthquakes, including time-domain and joint time–frequency-domain problems like static and dynamic triaxial test, acceleration time history, and Hilbert spectrum of rock and soil mass, which could study on the occurrence mechanism of earthquake inducing landslides more deeply and comprehensively.

6.2.3.1 Stress–Strain Relations of Sliding Mass Structural Surface

In order to carry out more in-depth study on landslide mechanism of bedrock and overburden slopes under seismic effects, this chapter, respectively, sets up stress Test Points N1–N5 and strain Test Points F1–F6 on sliding mass structural surface of single- and double-sided bedrock and overburden slopes, so as to analyze the failure mechanism of sliding mass structural surface under seismic effects from the perspective of stress and strain. Because the sliding process of single- and double-sided bedrock and overburden slopes is basically the same, so this chapter selects representative Test Point F3 of stress and strain on double-sided sliding mass structural surface as analyzing object. See Figs. 6.7 and 6.8 for detailed results. It is noteworthy that Fig. 6.8 shows that Test Point F3 has come across shear failure at T = 12.0, so this chapter only selects seismic response before 12.0 s to analyze.

Because all the mechanic indexes of soil mass decrease under seismic effects, tangent shear strength of sliding mass also decreases. Therefore, this chapter takes the strength index of soil mass under its own weight as basis to partially consider the safety. In the process of working out the strength, the following hypothesis was made: The soil mass is the ideal elastoplastic model and completely obeys M-C

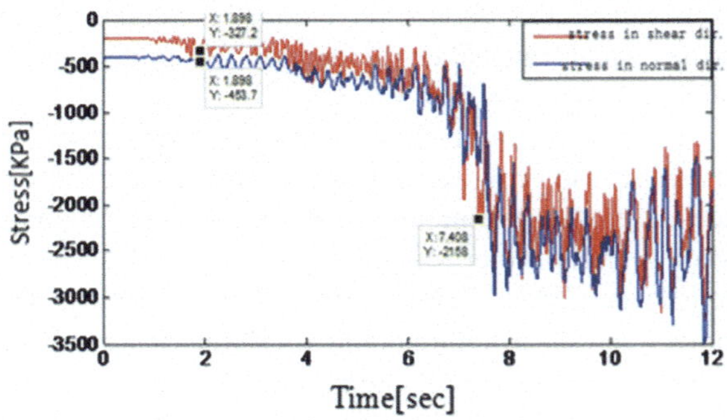

Fig. 6.7 Normal and tangent stress time history curve

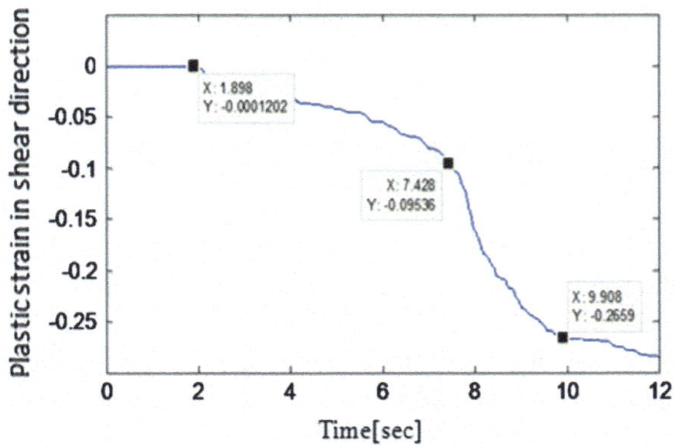

Fig. 6.8 Tangent plastic strain time history curve

failure principle. The lower bedrock is incompressible; that is, the normal compressive strength of structural surface is infinite. The tangent shear strength formula of sliding mass surface is obtained through calculation (see Formula 6.1). At the same time, due to the fact that GDEM could only adopt stress and strain data in X and Y directions when carrying out numerical calculation, this chapter will change coordinates according to Formulas (6.2) and (6.3).

$$P_\tau = C + \rho gh \cdot \cos\theta \cdot \tan\varphi \tag{6.1}$$

$$P_n' = T_x \cdot sn\theta + T_y \cos\theta \tag{6.2}$$

$$P_\tau' = T_y \cdot \sin\theta - T_x \cos\theta \tag{6.3}$$

In the above formulas, P_τ is the tangent shear strength of sliding mass surface, C stands for cohesive force of structural surface, φ stands for its internal friction angle, h represents the vertical thickness of upper covering layer of test points, θ is the bevel angle of slope surface and the horizontal surface, ρ is the density of upper covering layer, P_n' and P_τ', respectively, represent normal and shear stress and strain of sliding mass structural surface, and T_τ and T_y, respectively, stand for stress and strain in X and Y directions on sliding mass structural surface.

Bringing the relevant parameters into the above formulas, we will get that P_τ equals 2322 kPa, the shear and normal stress time history curve of sliding mass structural surface see Fig. 6.7. Through comprehensive analysis of Fig. 6.7, it can be known that under seismic ground motion effects with duration T = 12.0 s, the normal stress of test points has always been negative, which demonstrates that the structural surface has always been in compressed state and could not come across tensile failure. Therefore, this chapter will not consider the normal plastic strain time history curve of test points. See Fig. 6.8 for calculation results of plastic strain.

Through comprehensive analysis of Figs. 6.7 to 6.8, it can be known that before input seismic power T = 1.898 s, the tangent stress of test points is small, and the corresponding tangent plastic strain is almost 0, which shows that test points are still in elastic stage when T = 0–1.898 s; when T = 1.898–7.408 s, the tangent stress of test points gradually increases, so is tangent plastic stress, and the growth rate is on the rise; when T = 7.408 s, the tangent stress of test points is over its tangent shear strength for the first time, and the corresponding tangent plastic strain comes across the turning point at T = 7.428 s and suddenly increases afterward. At T = 7.408–12.0 s, the tangent stress of test points fluctuates around tangent shear strength line, and the tangent plastic stress time history curve comes across turning point at T = 9.908 s, with tangent plastic strain staying at 0.26 afterward. At the same time, it can be known from the characteristics of brittle fracture constitutive model and M-C failure principle that the test point has already been damaged, which is basically identical with the failure model of this test point in Fig. 6.6.

6.2.3.2 Vibration Acceleration Difference of Two Sides of Sliding Mass Structural Surface

In order to study the seismic wave propagating difference on sliding bed and inside the sliding mass, this chapter, respectively, sets up acceleration test points L1–L5 and C1–C6 inside the slope mass of single- and double-sided bedrock and overburden slopes and acceleration test points M1–M5 and D1–D6 inside the sliding mass to observe the propagating characteristics of seismic wave on the two sides of structural surface. Because the sliding process of single- and double-sided bedrock and overburden slopes is basically the same, this chapter selects representative Test Points (C3, D3) as analyzing objects. See Figs. 6.9 and 6.10 for detailed results.

Through comprehensive analysis of Figs. 6.9 and 6.10, it can be known that there exists big difference of horizontal acceleration of the two sides of sliding mass structural surface. Firstly, when seismic ground motion acceleration is small, the horizontal acceleration inside the sliding mass is over that inside the sliding bed; with the increase in seismic ground motion acceleration, the horizontal acceleration inside the sliding bed gradually surpasses that inside the sliding mass; secondly, there is a difference in the time of the two sides of sliding mass structural surface reaching the horizontal PGA with sliding mass slightly lagging behind. The reason for the above phenomenon may be that at the beginning of the earthquake, the seismic ground motion acceleration is small, few discontinuous shear failure points

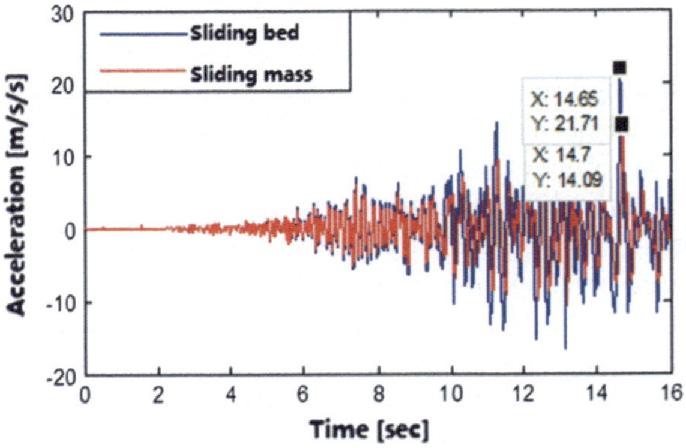

Fig. 6.9 Time history curve of horizontal acceleration

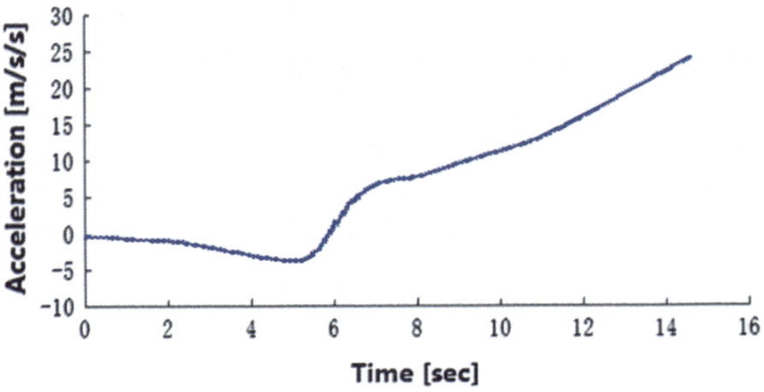

Fig. 6.10 Time history envelop of horizontal acceleration difference

occur on the structural surface, and the soil mass on the two sides of sliding mass structural surface did not come across relative slip. When seismic wave reaches the sliding mass structural surface, the PGA amplification effects occur inside the sliding mass. Based on the propagating theory of seismic wave in semi-infinite space, this chapter believes that reason for PGA amplification effects occurring inside the sliding mass may be as follows: Firstly, when seismic wave propagates from bedrock(sliding bed) to upper cover layer(sliding mass), the reflected wave comes across wave field separating phenomenon; for example, P wave reflects into P wave and SV wave and SV wave reflects into SV wave and P wave, and reflected wave of all wave types overlap together to form complex seismic wave field, which enlarges the motion of upper cover layer; secondly, the energy of seismic wave is mainly provided by P wave, SH wave and SV wave, and they can all enlarge the motion of upper layer when spreading from bedrock to upper cover layer. With the increase in seismic ground motion intensity, the acceleration amplification effects become more obvious, and acceleration difference of the two sides of structural surface also increases, while that will cause more shear failure points on the sliding mass structural surface, and they gradually connect together, forming shear failure surface. Because the soil mass on the two sides of structural surface comes across relative slip, the above theory loses efficacy, and at the same time, this phenomenon causes seismic wave energy dissipation at this position. With the increase of slippage, there will be more energy dissipation which causes sharp decreases in acceleration inside the sliding mass and leads to the acceleration inside the sliding bed exceeding that inside the sliding mass.

6.2.3.3 Dynamic Characteristics of Soil Particles Inside Sliding Mass

In order to study the dynamic characteristics of soil particles inside the sliding mass under the effects of dynamic loading, this chapter carries out dynamic triaxial test focusing on the sliding mass materials on the high and steep prototype slope and obtains the shear strength of soil particles inside the sliding mass under different seismic intensity. See Figs. 6.11 and 6.2 for detailed results. It is noteworthy that because the dynamic response rules of sliding mass materials of single- and double-sided bedrock and overburden slopes are basically the same under effects of dynamic circulative loading.

Through comprehensive analysis of Table 6.2, it can be known that in the same vibration cycle, the growth amplitude of dynamic shear strength τ_d gradually increases with rising confining pressure; under the same confining pressure, the dynamic shear strength τ_d gradually decreases with the increase in vibration cycle N, i.e., increase in correspondent seismic grade. Therefore, the shear strength inside the sliding bed and sliding mass structural surface will gradually decrease with the growth of seismic ground motion acceleration.

Fig. 6.11 Dynamic triaxial test of saturated sample

Table 6.2 Table of dynamic shear strength test results of sliding mass materials

Vibrating frequency N		12	20	30
Correspondent level		7	7.5	8
		Cyclic stress ratio τ_d/σ_{Sc}		
Confining pressure σ_{Sc}	100	0.407	0.382	0.362
	200	0.395	0.396	0.35
	400	0.373	0.335	0.322

6.2.3.4 Inconsistency of Motion Between Sliding Bed and Sliding Mass

In order to study the inconsistency of motion between sliding bed and sliding mass, this chapter, respectively, sets up acceleration test points L1–L5 and C1–C6 inside the slope mass of single- and double-sided high and steep bedrock and overburden slopes and acceleration test points M1–M5 and D1–D6 inside the sliding mass. Because the sliding process of single- and double-sided bedrock and overburden slopes is basically the same, this chapter selects representative Test Points (C3, D3) on the two sides of sliding mass structural surface of double-sided bedrock and overburden slope as analyzing objects. The inconsistency of motion between sliding bed and sliding mass mainly involves frequency, amplitude, and time of them, and the current time-domain analysis and frequency-domain analysis could only conduct time–amplitude or frequency–amplitude research, but could not organically combine amplitude, frequency, and time together, so this chapter will adopt Hilbert–Huang transform and select horizontal acceleration time history curve of C3 and D3 to study sliding bed and sliding mass motion inconsistency from the perspective of joint time–frequency domain (see Figs. 6.12 and 6.13).

First, explain the definition of instant frequency adopted by the dissertation according to Hu Yuxian et al.: The instant frequency $\omega(t)$ at any time is frequency ω to make instant Hilbert spectrum value $H(t, \omega)$ of this moment reaching its maximum, and that is the dominant frequency of the instant Hilbert spectrum value $H(t, \omega)$ of this moment. The changing process of this instant dominant frequency is the instant frequency process of initial acceleration record.

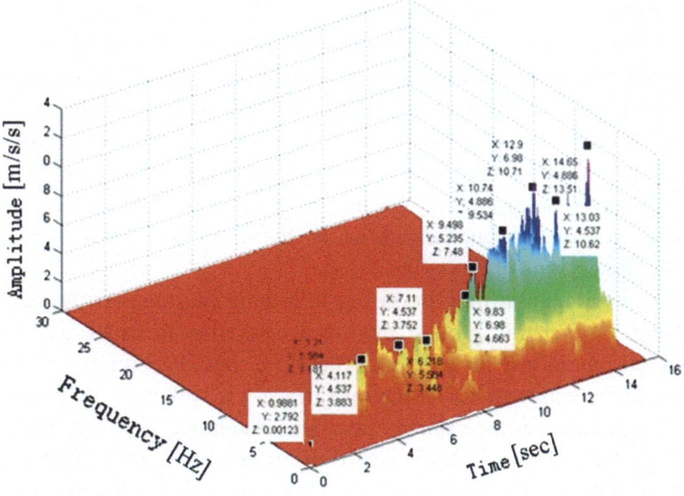

Fig. 6.12 Hilbert spectrum of horizontal acceleration inside the sliding bed

Fig. 6.13 Hilbert spectrum of horizontal acceleration inside the sliding mass

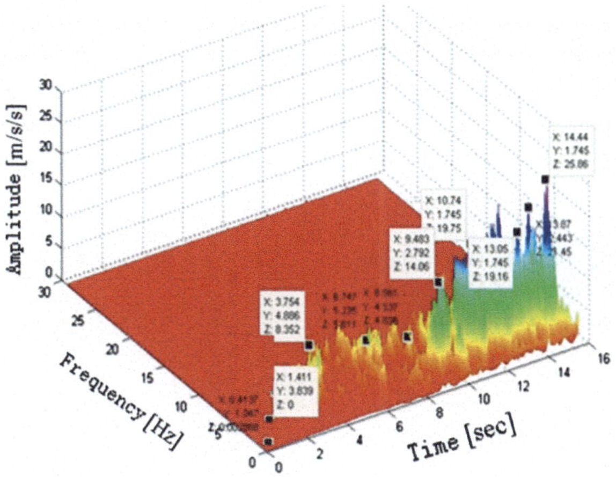

It can be known from Figs. 6.12 and 6.13 that in the horizontal acceleration, Hilbert spectrum of the two sides of sliding mass structural surface exists certain similarities and differences. **Similarities**: the horizontal acceleration of the two sides of sliding mass structural surface mainly concentrate in 0–10 Hz, and although some acceleration exist in 10–30 Hz, but they are with small amplitude, which is identical with the calculation results of Fourier spectrum. **Differences**: first, at the beginning of earthquake, the amplitude of horizontal acceleration inside the sliding mass is over that inside the sliding bed; with the increase of seismic intensity, the amplitude of horizontal acceleration inside the sliding bed surpassing that inside the sliding mass; second, based on the previous definition of instant frequency, select the acceleration amplitude in Hilbert spectrum at any time and find out its correspondent frequency to draw the time–frequency time history curve inside the sliding bed and sliding mass (see Figs. 6.14 and 6.15). It can be known from Figs. 6.14 to 6.15 that when seismic ground motion is small, the instant frequency of horizontal acceleration inside the sliding bed and sliding mass all stabilizes in a certain range, with the former slightly higher than the latter; with the increase in seismic ground motion, the instant frequency of horizontal acceleration inside the sliding bed stays unchanged, while that inside the sliding mass gradually decreases and finally becomes stable.

The above phenomenon may lie in the following reasons: At the beginning of the earthquake, the seismic ground motion acceleration is small, and the sliding mass surface has enough shear strength. Therefore, few discontinuous shear failure points appear at the sliding mass structural surface, the soil mass on the two sides of sliding mass structural surface don't come across relative slip, and the motion of sliding bed and sliding mass are basically the same which causes the instant frequency of horizontal acceleration inside the sliding bed and sliding mass stabilize in a certain range with the former slightly higher than the latter. When seismic wave reaches the sliding mass structural surface, the wave impedance of deposit is smaller than that of the sliding bed, which causes horizontal acceleration coming across amplification effects inside the sliding mass. With the increase in seismic ground motion intensity, the acceleration amplification effects become more obvious, and acceleration difference of the two sides of structural surface also increases, while the latter will cause more shear failure points on the sliding mass structural surface, and they gradually connect together, forming shear failure

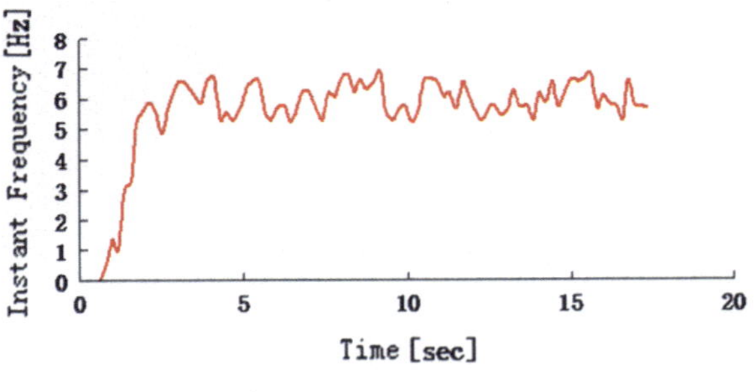

Fig. 6.14 Instant frequency of horizontal acceleration inside the sliding bed

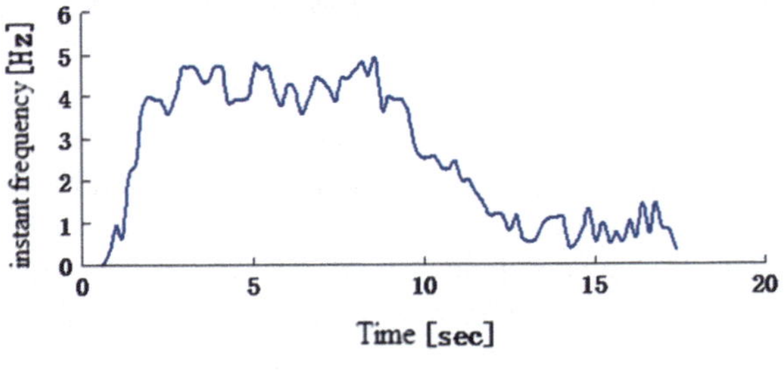

Fig. 6.15 Instant frequency of horizontal acceleration inside the sliding mass

surface. The soil mass of the two sides of structural surface comes across relative slip, the instant frequency of horizontal acceleration inside the sliding mass gradually decreases, and the seismic wave dissipates at this position. With the increase in slippage, there will be more energy dissipation which causes sharp decreases in acceleration inside the sliding mass and leads to the acceleration inside the sliding bed exceeding that inside the sliding mass. The instant frequency of horizontal acceleration inside the sliding mass also stabilizes in a certain range.

It can be known from the above general analysis that at the beginning of the earthquake, the seismic ground motion acceleration is small, and the motion between sliding bed and sliding mass has strong consistency. With the increase in seismic ground motion strength, the difference of vibration frequency and amplitude between sliding bed and sliding mass gradually enlarges, and the motion inconsistency gradually strengthens, while the strengthening of motion inconsistency will intensify the development of tensile and shear failure points on structural surface and lead to landslides. Therefore, the inconsistency of motion between sliding bed and sliding mass is a main controlling factor leading to landslides of bedrock and overburden slopes.

6.2.3.5 Difference of Seismic Energy Distribution and Dissipation Between Sliding Bed and Sliding Mass

In order to study the difference of seismic energy distribution and dissipation between sliding bed and sliding mass, this chapter, respectively, sets up acceleration test points L1–L5 and C1–C6 inside the slope mass of single- and double-sided high and steep bedrock and overburden slopes and acceleration test points M1–M5 and D1–D6 inside the sliding mass. Because the sliding process of single- and double-sided bedrock and overburden slopes is basically the same, this chapter selects representative Test Points (C1–C6, D1–D6) on the two sides of sliding mass structural surface of double-sided bedrock and overburden slope as analyzing objects. At the same time, the seismic energy inside the sliding mass mainly comes from transmission energy of seismic wave inside sliding bed passing through the sliding structural surface. Therefore, the transmission coefficient of sliding mass structural surface could well reflect the difference of seismic ground motion energy between sliding bed and sliding mass. Therefore, this chapter will use Hilbert–Huang transform and work out the energy distribution of all test points (C1–C6, D1–D6) from the joint time–frequency domain so as to obtain the time–frequency curve of transmission coefficient of the whole sliding mass structural surface (see Fig. 6.16). It is noteworthy that the spectrum values in the Fourier spectrum only stand for the amount of energy and have no physical meaning. This chapter defines the transmission coefficient as follows: energy transmission coefficient = energy inside the sliding mass/energy inside the sliding bed.

It can be known from Fig. 6.16 that there is a big difference of energy distribution between sliding bed and sliding mass. First, when the seismic motion is small, the energy transmission coefficient of seismic motion inside the sliding mass

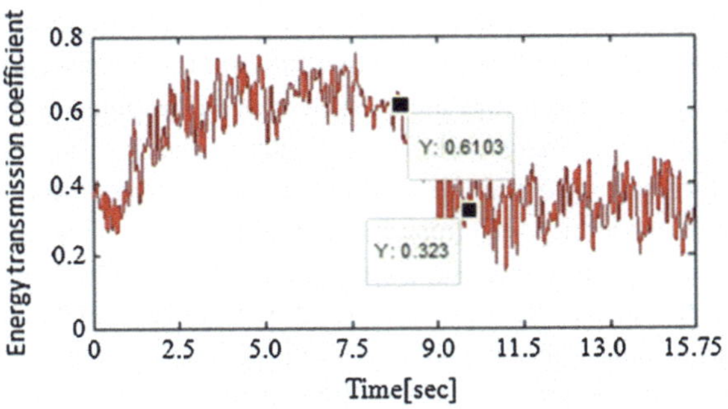

Fig. 6.16 Energy transmission coefficient time history curve of sliding mass structural surface

stabilizes at 0.55–0.75; with the increase of seismic intensity, it gradually decreases and stabilizes at 0.20–0.40. According to Figs. 6.12 and 6.13, it can be known that the seismic motion energy inside the sliding bed mainly is distributed in the dominant frequency range of 5–7 Hz, while that inside the sliding mass mainly is distributed in the dominant frequency ranges of 0.5–1.5 Hz and 4–5 Hz, with the lower dominant frequency range controlling the major power of seismic wave. Therefore, with the increase in seismic intensity, the energy transmission coefficient decreases, with the controlling frequency range of seismic motion energy inside the sliding mass transferring from high-frequency range to low-frequency range.

The reason for the above phenomenon may be that when seismic ground motion is small, there is no relative slip between sliding bed and sliding mass, the instant frequency stabilizes at 5–7 Hz, and the energy transmission coefficient is high, stabilizing in the range of 0.55–0.75. At the same time, using elastic wave theory can get the calculation result 0.665 of energy transmission coefficient. These two are basically of the same situation. Therefore, at the beginning of the earthquake, the energy of seismic ground motion inside the sliding bed mainly concentrates in the range of 5–7 Hz, and that of the sliding mass mainly concentrates in the range of 4–5 Hz.

With the increase in seismic intensity, the energy of input seismic wave gradually rising and the shear and tensile failure points on the sliding mass structural surface also becomes more, which gradually forms the connecting failure surface. Relative slip occurs between sliding bed and sliding mass, which causes the increase in energy dissipation of seismic wave on the structural surface, leading to the decrease of energy transmission coefficient, which finally stabilizes in the range of 0.20–0.40. At the same time, with the appearance of inconsistency between sliding bed and sliding mass, the instant frequency of seismic acceleration inside the sliding mass gradually decreases and stabilizes in the range of 0.5–1.5 Hz. Thus, after relative slip between sliding bed and sliding mass, the controlling frequency range of seismic motion energy inside the sliding mass gradually stabilizes in the range of 0.5–1.5 Hz. Therefore, with the increase in seismic intensity, the damage of sliding mass structural surface enlarges, the relative slippage between the sliding bed and sliding mass is also on the rise, the energy dissipation on the structural surface become more, energy transmission coefficient decreases and the controlling frequency range of seismic motion energy inside the sliding mass transferring from high-frequency range to low-frequency range.

Through comprehensive analysis above, it can be known that at the beginning of seismic ground motion, energy transmission coefficient of sliding mass structural surface is high and large amount of seismic motion energy is stored inside the cover layer, while the thickness of the cover layer is thin, which causes the energy intensity of storage; dissipation inside the cover layer is much more than that in the lower sliding bed, leading to the tensile and shear failure of sliding mass structural surface. With the increase in seismic intensity, the difference of energy between sliding bed and sliding mass gradually enlarges, which intensifies the damage of sliding mass structural surface and leads to landslides. Thus, difference of seismic energy distribution and dissipation between sliding bed and sliding mass is a key factor to induce landslides, which can fully explain the seismic damage phenomenon of "peeling-type" landslides widely spread in the soft rock distribution region and hard rock superficial weathered zone in Wenchuan earthquake.

6.3 Brief Summary

This chapter focused on the two major landslide hazard spots on the left side of G213, built the numerical analysis model based on the shaking table test model, and used the new numerical calculation software—GDEM to simulate the whole process of deposit body landslides of single- and double-sided bedrock and overburden slopes from deformation to failure slip under intense earthquake, conducting comprehensive and in-depth research of slope deformation features and formation mechanism of bedrock and overburden slopes from the perspective of time domain and joint time–frequency domain, and came to the following conclusions:

(1) No matter for single- or double-sided bedrock and overburden slopes, the landslide processes are basically the same, i.e., under effects of gravity and seismic power, tensile stress concentration firstly occurs at sliding mass top, causing deformation of sliding mass along sliding mass structural surface trailing edge, which leads to tensile and shear failure points at this position. Then, with continuous seismic power, the shear failure points on sliding mass structural surface gradually develop to lock-fixed section of sliding mass leading edge with increasing amount of tensile failure points on sliding mass surface, which finally causes progressive damage of lock-fixed section, crack surface connecting to sliding zone, and sliding mass slipping from shear crack into landslide.

(2) The inconsistency of motion between sliding bed and sliding mass, difference of seismic energy distribution and dissipation, and decrease of sliding mass structural surface strength with the increase of seismic intensity are three major controlling factors inducing landslides;

(3) When seismic ground motion acceleration is small, the instant frequency of horizontal acceleration inside the sliding bed and sliding mass stabilizes in a certain range with the former slightly higher than the latter. At the same time, the energy transmission coefficient on the sliding mass structural surface and the controlling frequency range of seismic energy inside the sliding mass all stabilizes in a certain range; with the increase in seismic intensity, the instant frequency of acceleration inside the sliding mass gradually decreases, and the energy transmission coefficient of sliding mass structural surface gradually decreases, with both finally approaching stability. Simultaneously, the controlling frequency range of seismic motion energy inside the sliding mass transferring from high-frequency range to low-frequency range.

Chapter 7
Slope Seismic Stability

The seismic stability of slopes has been the hot spot issues of geological disaster study, and especially after the 5.12 Wenchuan earthquake, it has become the study focus of specialists and scholars. For slope seismic stability analysis, the current method includes pseudo-static method, sliding block analysis method, and numerical simulation, in which pseudo-static method only takes consideration of the influence of PGA on slope stability, but not the influence of frequency and duration of seismic waves; neither sliding block analysis method can solve the instant stability coefficient of slopes under seismic loading effects, nor can it accurately evaluate the seismic stability in real engineering, so it is not convenient to use; although numerical simulation can carry out time–frequency analysis of rock slope stability under seismic effects, the operation is complex, especially for high and steep slopes with large difference in rock layer distribution. However, the results of Chap. 6 show that bedrock and overburden slopes coming across landslides are induced by duration, frequency, and amplitude of seismic ground motion, and these three form an organic whole, so singly considering one of them will lead to partial recognition of slope seismic stability. Therefore, it is quite necessary to study the influence of time, frequency, and amplitude on slope seismic stability time–frequency analysis.

According to the field research results and formation mechanism analysis results of large-scale landslide disasters in Wenchuan earthquake, it can be known that SV wave (elastic wave whose vibration direction and propagation direction are vertical to the slope cross section) is the most disastrous and causes the most serious damage. In addition, the major energy of seismic wave is controlled by SV wave, while the duration of P wave is short and its energy is small. Therefore, we can approximately take it that the slope stability under earthquake is the slope stability under SV wave. But SV wave is a complex non-stationary signal, whose frequency feature changes with time. At the same time, bedrock and overburden slope are the most common slope structure with maximum instability quantity in seismic areas. Therefore, this chapter will focus on bedrock and overburden slope, based on elastic wave theory, adopt Hilbert–Huang transform and general analysis model of slopes,

Y. Changwei et al., *Slope Earthquake Stability*,
DOI 10.1007/978-981-10-2380-4_7

and propose time–frequency analysis method of bedrock and overburden slope seismic stability which could take consideration of three factors of seismic waves so as to provide new ideas to bedrock and overburden slope stability recognition.

7.1 Slope Seismic Stability

Based on elastic wave theory and general geological analysis model, we adopt Hilbert–Huang transform, propose time–frequency analysis method of bedrock and overburden slope seismic stability, and verify its rationality through shaking table test and numerical analysis results.

7.1.1 General Thinking and Basic Assumption

The time–frequency analysis method derivation of bedrock and overburden slopes mainly follows the following process: Generalize the analysis model, make sliding mass structural surface meshing, and adopt HHT and elastic wave theory to solve the instant stress components due to reflection and transmission when seismic wave passes through this point; take consideration of gravitational stress field, use the M-C failure principle, discuss its damage, and work out the stress component of other meshing points in turn; and draw the instant stress distribution of each point on the structural surface and finally work out the instant stability coefficient of bedrock and overburden slopes under earthquake. The derivation of this method will be based on the following hypothesis: ① The closely connected medium on the two sides of sliding mass structural surface is homogeneous and isotropic elastomer; ② the sliding mass structural surface conforms to the M-C failure principle; ③ due to the fact that the seismic waves continuously coming across refraction and transmission when transmitting from seismic source upward, which causes them close to the vertical direction when reaching the earth surface, this chapter selects seismic waves transmitting vertically upward from bedrock to slope mass as input wave; and ④ because wave-type conversion will occur when seismic wave arrives at the cover layer surface with its major energy transferring into Rayleigh surface wave moving along the cover layer surface, and the material damping inside the cover layer will largely consume the seismic wave energy, which will finally lead to reflective seismic wave on the cover layer surface bringing very weak seismic energy to the sliding mass structural surface, this chapter will pay no attention to the influence of reflective seismic wave on cover layer surface to sliding mass structural surface.

7.1.2 Generalized Analysis Model

It can be known from the research results of Wenchuan earthquake hazard that bedrock and overburden slope are the most common slope structure with maximum instability quantity in seismic areas, and a common feature of slopes of this kind is as follows: Obvious structural surface exists inside the slope mass, the material property on its two sides has certain difference, and structural surface is the major controlling factor of slope landslides. Therefore, this chapter builds the generalized analysis model based on the above thinking (see Fig. 7.1). Figure 7.1 shows the upper surface of the cover layer, I–I interface is of any shape, the tangential obliquity of I–I interface at micro-unit A is θ, and evenly mesh the sliding mass surface into 7 points (take this as example) and select A as study object. Figure 7.2 shows the reflective and transmission model of SV wave at Point A, the physico-mechanical parameters of the middle-upper medium of this model (bedrock) are, respectively, ρ_1, C_{p1}, C_{s1} and G_1 (density, P wave speed, S wave speed, and shear modulus), and the physico-mechanical parameters of the lower medium are ρ_2, C_{p2}, C_{s2} and G_2 (density, P wave speed, S wave speed, and shear modulus). In there, I–I interface is the sliding mass structural surface; when SV wave S_1 is transmitted onto the I–I interface, it will generate reflective SV wave S_2, reflective P wave S_3, transmission SV wave S_4, and transmission P wave S_5, while α_1, α'_1 β_1, and β'_1, respectively, represent the incident angle of SV wave, reflection angle of P

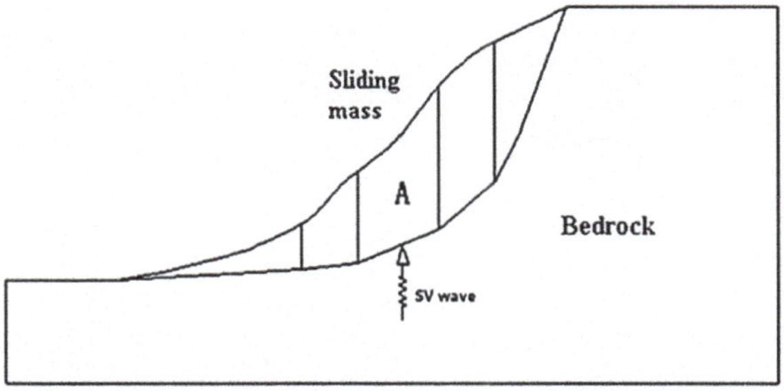

Fig. 7.1 Generalized model of bedrock and overburden slopes

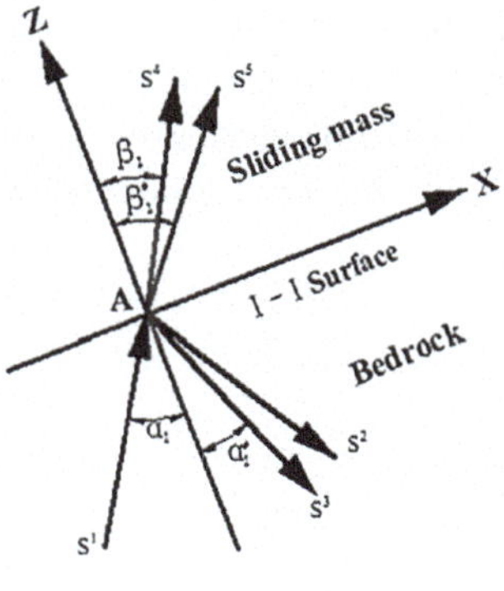

Fig. 7.2 Reflective and transmission analysis model of Point A

wave, and transmission angle of P wave. Note: the incident angle and reflection angle of SV wave are equal.

7.1.3 *Reflective and Transmission Coefficients*

In the *XOZ* plane in Fig. 7.2, the potential function of all the stress waves can be represented by elastic displacement amplitude, and the potential function of incident S wave, reflected S wave, reflected P wave, transmitted S wave, and transmitted P wave could be, respectively, written into Formula (7.1) to (7.4).

$$\mathrm{S}^i = \mathrm{S}_0^i e^{j(k_x^{i1}X + k_z^{i1}Z - \omega t)}, \quad i = 1,2,3,4,5 \tag{7.1}$$

In the formulas, $i = 1,2,3,4,5$, respectively, stand for incident S wave S^1, reflected S wave S^2, reflected P wave S^3, transmitted S wave S^4, and transmitted P wave S^5; k_x^{i1} and k_z^{i1}, respectively, represent wave vector of incident wave, reflected wave, and transmitted wave along *X* and *Z* directions, according to Snell theorem $k_x^{11} = k_x^{21} = k_x^{31} = k_x^{41} = k_x^{51}$.

$$k_x^{11} = \frac{\omega}{V_{\mathrm{s1}}}\sin\alpha_1; k_x^{21} = \frac{\omega}{V_{\mathrm{s1}}}\sin\alpha_1; k_x^{31} = \frac{\omega}{V_{\mathrm{p1}}}\sin\alpha_1'; k_x^{41} = \frac{\omega}{V_{\mathrm{s2}}}\sin\beta_1; k_x^{51} = \frac{\omega}{V_{\mathrm{p2}}}\sin\beta_1' \tag{7.2}$$

$$k_z^{11} = \frac{\omega}{V_{\mathrm{s1}}}\cos\alpha_1; k_z^{21} = -\frac{\omega}{V_{\mathrm{s1}}}\cos\alpha_1; k_z^{31} = -\frac{\omega}{V_{\mathrm{p2}}}\cos\alpha_1'; k_z^{41} = \frac{\omega}{V_{\mathrm{s2}}}\cos\beta_1; k_z^{51} = \frac{\omega}{V_{\mathrm{p2}}}\cos\beta_1' \tag{7.3}$$

$$\frac{\mathrm{Sin}\alpha_1}{V_{s1}} = \frac{\mathrm{Sin}\alpha_1'}{V_{p1}} = \frac{\mathrm{Sin}\beta_1}{V_{s2}} = \frac{\mathrm{Sin}\beta_1'}{V_{p2}} \tag{7.4}$$

in the formulas, V_{p1}, V_{s1}, respectively, stand for the speed of S wave and P wave in the upper medium; V_{p2}, V_{s2}, respectively, stand for the speed of S wave and P wave in the lower medium.

When incident wave is on the *XOY* plane, i.e., when taking effects at Z = 0, it should meet the following conditions:

$$u_{1z}(x,z,t) = u_{2z}(x,z,t); u_{1x}(x,z,t) = u_{2x}(x,z,t); \sigma_{1z}(x,z,t) = \sigma_{2z}(x,z,t); \tau_{1z}(x,z,t) = \tau_{2z}(x,z,t)$$

It can be known from Fig. 7.2 that in the upper medium, the horizontal and vertical displacement components are as follows:

$$\begin{cases} u_1 = \mathrm{S}^1 \cos\alpha_1 - \mathrm{S}^2 \cos\alpha_1 + \mathrm{S}^3 \sin\alpha_1' \\ \omega_1 = -\mathrm{S}^1 \sin\alpha_1 - \mathrm{S}^2 \sin\alpha_1 - \mathrm{S}^3 \cos\alpha_1' \end{cases} \tag{7.5}$$

In the lower medium, the horizontal and vertical displacement components are as follows:

$$\begin{cases} u_2 = \mathrm{S}^4 \cos\beta_1 + \mathrm{S}^5 \sin\beta_1' \\ \omega_2 = -\mathrm{S}^4 \sin\beta_1 + \mathrm{S}^5 \cos\beta_1' \end{cases} \tag{7.6}$$

Solving simultaneously with Formula (7.5) to (7.6) can get displacement boundary condition:

$$\mathrm{S}^1 \cos\alpha_1 - \mathrm{S}^2 \cos\alpha_1 + \mathrm{S}^3 \sin\alpha_1' = \mathrm{S}^4 \cos\beta_1 + \mathrm{S}^5 \sin\beta_1' \tag{7.7}$$

$$-\mathrm{S}^1 \sin\alpha_1 - \mathrm{S}^2 \sin\alpha_1 - \mathrm{S}^3 \cos\alpha_1' = -\mathrm{S}^4 \sin\beta_1 + \mathrm{S}^5 \cos\beta_1' \tag{7.8}$$

Bringing the relevant parameters into Formulas (7.7) and (7.8), after simplification, we can get:

$$\mathrm{S}_0^1 \cos\alpha_1 - \mathrm{S}_0^2 \cos\alpha_1 + \mathrm{S}_0^3 \sin\alpha_1' = \mathrm{S}_0^4 \cos\beta_1 + \mathrm{S}_0^5 \sin\beta_1' \tag{7.9}$$

$$-\mathrm{S}_0^1 \sin\alpha_1 - \mathrm{S}_0^2 \sin\alpha_1 - \mathrm{S}_0^3 \cos\alpha_1' = -\mathrm{S}_0^4 \sin\beta_1 + \mathrm{S}_0^5 \cos\beta_1' \tag{7.10}$$

Calculating the stress on I–I interface with isotropic Hooke's law:

$$\begin{cases} \sigma_z = \lambda\left(\frac{\partial u_x}{\partial x} + \frac{\partial u_z}{\partial z}\right) + 2G\frac{\partial u_z}{\partial z} \\ \tau_{zx} = G\left(\frac{\partial u_x}{\partial z} + \frac{\partial u_z}{\partial x}\right) \end{cases} \tag{7.11}$$

In the formula, λ stands for Lame constant and G stands for shear modulus.

Solving simultaneously with Formulas (7.1)–(7.6) and Formula (7.11), we can get stress boundary condition:

$$\begin{aligned} &\lambda_1[\mathrm{S}_0^1 \cdot k_x^{11} \cdot \cos\alpha_1 - \mathrm{S}_0^2 \cdot k_x^{21} \cdot \cos\alpha_1 + \mathrm{S}_0^3 \cdot k_x^{31} \cdot \sin\alpha_1'] + (\lambda_1 + 2\mu_1)[-\mathrm{S}_0^1 \cdot k_z^{11} \cdot \sin\alpha_1 \\ &- \mathrm{S}_0^2 \cdot k_z^{21} \cdot \sin\alpha_1 - \mathrm{S}_0^3 \cdot k_z^{31} \cdot \cos\alpha_1'] = \lambda_2[\mathrm{S}_0^4 \cdot k_x^{41} \cdot \cos\beta_1 + \mathrm{S}_0^5 \cdot k_x^{51} \cdot \sin\beta_1'] + (\lambda_2 + 2\mu_2) \\ &[-\mathrm{S}_0^4 \cdot k_z^{41} \cdot \sin\beta_1 + \mathrm{S}_0^5 \cdot k_z^{51} \cdot \cos\beta_1'] \end{aligned} \tag{7.12}$$

$$\mu_1[S_0^1 \cdot k_z^{11} \cdot \cos\alpha_1 - S_0^2 \cdot k_z^{21} \cdot \cos\alpha_1 + S_0^3 \cdot k_z^{31} \cdot \sin\alpha_1' - S_0^1 \cdot k_x^{11} \cdot \sin\alpha_1 - S_0^2 \cdot k_x^{21} \cdot \sin\alpha_1 - S_0^3 \cdot k_x^{31} \cdot \cos\alpha_1'] = \mu_2[S_0^4 \cdot k_z^{41} \cdot \cos\beta_1 + S_0^5 \cdot k_z^{51} \cdot \sin\beta_1' + S_0^5 \cdot k_x^{51} \cdot \cos\beta_1' - S_0^4 \cdot k_x^{41} \cdot \sin\beta_1] \tag{7.13}$$

Through comprehensive analysis of Formulas (7.10)–(7.13), we can get:

$$\begin{cases} S_0^1\cos\alpha_1 - S_0^2\cos\alpha_1 + S_0^3\sin\alpha_1' = S_0^4\cos\beta_1 + S_0^5\sin\beta_1' \\ -S_0^1\sin\alpha_1 - S_0^2\sin\alpha_1 - S_0^3\cos\alpha_1' = -S_0^4\sin\beta_1 + S_0^5\cos\beta_1' \\ \lambda_1[S_0^1 \cdot k_x^{11} \cdot \cos\alpha_1 - S_0^2 \cdot k_x^{21} \cdot \cos\alpha_1 + S_0^3 \cdot k_x^{31} \cdot \sin\alpha_1'] + (\lambda_1 + 2\mu_1)[-S_0^1 \cdot k_z^{11} \cdot \sin\alpha_1 \\ -S_0^2 \cdot k_z^{21} \cdot \sin\alpha_1 - S_0^3 \cdot k_z^{31} \cdot \cos\alpha_1'] = \lambda_2[S_0^4 \cdot k_x^{41} \cdot \cos\beta_1 + S_0^5 \cdot k_x^{51} \cdot \sin\beta_1'] + (\lambda_2 + 2\mu_2) \\ [-S_0^4 \cdot k_z^{41} \cdot \sin\beta_1 + S_0^5 \cdot k_z^{51} \cdot \cos\beta_1'] \\ \mu_1[S_0^1 \cdot k_z^{11} \cdot \cos\alpha_1 - S_0^2 \cdot k_z^{21} \cdot \cos\alpha_1 + S_0^3 \cdot k_z^{31} \cdot \sin\alpha_1' - S_0^1 \cdot k_x^{11} \cdot \sin\alpha_1 - S_0^2 \cdot k_x^{21} \cdot \sin\alpha_1 \\ -S_0^3 \cdot k_x^{31} \cdot \cos\alpha_1'] = \mu_2[S_0^4 \cdot k_z^{41} \cdot \cos\beta_1 + S_0^5 \cdot k_z^{51} \cdot \sin\beta_1' + S_0^5 \cdot k_x^{51} \cdot \cos\beta_1' - S_0^4 \cdot k_x^{41} \cdot \sin\beta_1] \end{cases} \tag{7.14}$$

After simplifying Formula (7.14), we can get:

$$\begin{cases} S_0^1\cos\alpha_1 - S_0^2\cos\alpha_1 + S_0^3\sin\alpha_1' - S_0^4\cos\beta_1 - S_0^5\sin\beta_1' = 0 \\ -S_0^1\sin\alpha_1 - S_0^2\sin\alpha_1 - S_0^3\cos\alpha_1' + S_0^4\sin\beta_1 - S_0^5\cos\beta_1' = 0 \\ [\lambda_1 \cdot k_x^{11} \cdot \cos\alpha_1 - (\lambda_1 + 2\mu_1) \cdot k_z^{11} \cdot \sin\alpha_1]S_0^1 - [\lambda_1 \cdot k_x^{21} \cdot \cos\alpha_1 + (\lambda_1 + 2\mu_1) \cdot k_z^{21} \\ \cdot \sin\alpha_1]S_0^2 + [\lambda_1 \cdot k_x^{31} \cdot \sin\alpha_1' - (\lambda_1 + 2\mu_1) \cdot k_z^{31} \cdot \cos\alpha_1']S_0^3 + [(\lambda_2 + 2\mu_2) \cdot k_z^{41} \cdot \sin\beta_1 \\ -\lambda_2 \cdot k_x^{41} \cdot \cos\beta_1]S_0^4 - [\lambda_2 \cdot k_x^{51} \cdot \sin\beta_1' + (\lambda_2 + 2\mu_2) \cdot k_z^{51} \cdot \cos\beta_1']S_0^5 = 0 \\ \mu_1[k_z^{11} \cdot \cos\alpha_1 - k_x^{11} \cdot \sin\alpha_1]S_0^1 - \mu_1[k_z^{21} \cdot \cos\alpha_1 + k_x^{21} \cdot \sin\alpha_1]S_0^2 + \mu_1[k_z^{31} \cdot \sin\alpha_1' \\ -k_x^{31} \cdot \cos\alpha_1']S_0^3 - \mu_2[k_z^{41} \cdot \cos\beta_1 - k_x^{41} \cdot \sin\beta_1]S_0^4 - \mu_2[k_z^{51} \cdot \sin\beta_1' + k_x^{51} \cdot \cos\beta_1']S_0^5 = 0 \end{cases} \tag{7.15}$$

Dividing each item in Formula (7.15) with S_0^1 and making $A' = S_0^2/S_0^1, B' = S_0^3/S_0^1, C' = S_0^4/S_0^1, D' = S_0^5/S_0^1$, we can get transmitted and reflected coefficient equation set after rewriting:

$$\begin{cases} \cos\alpha_1 - A'\cos\alpha_1 + B'\sin\alpha_1' - C'\cos\beta_1 - D'\sin\beta_1' = 0 \\ -\sin\alpha_1 - A'\sin\alpha_1 - B'\cos\alpha_1' + C'\sin\beta_1 - D'\cos\beta_1' = 0 \\ [\lambda_1 \cdot k_x^{11} \cdot \cos\alpha_1 - (\lambda_1 + 2\mu_1) \cdot k_z^{11} \cdot \sin\alpha_1] - [\lambda_1 \cdot k_x^{21} \cdot \cos\alpha_1 + (\lambda_1 + 2\mu_1) \cdot k_z^{21} \\ \cdot \sin\alpha_1]A' + [\lambda_1 \cdot k_x^{31} \cdot \sin\alpha_1' - (\lambda_1 + 2\mu_1) \cdot k_z^{31} \cdot \cos\alpha_1']B' + [(\lambda_2 + 2\mu_2) \cdot k_z^{41} \\ \cdot \sin\beta_1 - \lambda_2 \cdot k_x^{41} \cdot \cos\beta_1]C' - [\lambda_2 \cdot k_x^{51} \cdot \sin\beta_1' + (\lambda_2 + 2\mu_2) \cdot k_z^{51} \cdot \cos\beta_1']D' = 0 \\ \mu_1[k_z^{11} \cdot \cos\alpha_1 - k_x^{11} \cdot \sin\alpha_1] - \mu_1[k_z^{21} \cdot \cos\alpha_1 + k_x^{21} \cdot \sin\alpha_1]A' + \mu_1[k_z^{31} \\ \cdot \sin\alpha_1' - k_x^{31} \cdot \cos\alpha_1']B' - \mu_2[k_z^{41} \cdot \cos\beta_1 - k_x^{41} \cdot \sin\beta_1]C' - \mu_2[k_z^{51} \cdot \sin\beta_1' \\ + k_x^{51} \cdot \cos\beta_1']D' = 0 \end{cases} \tag{7.16}$$

Simplifying Formula (7.16) into matrix equation:

$$BX = c \Leftrightarrow [B_1 \quad B_2] \cdot \begin{bmatrix} A' \\ B' \\ C' \\ D' \end{bmatrix} = \begin{bmatrix} -\cos\alpha_1 \\ \sin\alpha_1 \\ -(\lambda_1 \cdot k_x^{11} \cdot \cos\alpha_1 - (\lambda_1 + 2\mu_1) \cdot k_z^{11} \cdot \sin\alpha_1) \\ -\mu_1 (k_z^{11} \cdot \cos\alpha_1 - k_x^{11} \cdot \sin\alpha_1) \end{bmatrix} \tag{7.17}$$

In Formula (7.17), the concrete forms of coefficients $B1$ and $B2$ are as follows:

$$B_1 = \begin{bmatrix} -\cos\alpha_1 & \sin\alpha_1' \\ -\sin\alpha_1 & -\cos\alpha_1' \\ -\lambda_1 \cdot k_x^{21}\cos\alpha_1 - (\lambda_1 + 2\mu_1)k_z^{21}\sin\alpha_1 & \lambda_1 \cdot k_x^{31}\sin\alpha_1' - (\lambda_1 + 2\mu_1) \cdot k_z^{31} \cdot \cos\alpha_1' \\ -\mu_1(k_z^{21}\cos\alpha_1 + k_x^{21}\sin\alpha_1) & \mu_1(k_z^{31}\sin\alpha_1' - k_x^{31}\cos\alpha_1') \end{bmatrix} \tag{7.18}$$

$$B_2 = \begin{bmatrix} -\cos\beta_1 & -\sin\beta_1' \\ \sin\beta_1 & -\cos\beta_1' \\ -\lambda_2 \cdot k_x^{41}\cos\beta_1 + (\lambda_2 + 2\mu_2)k_z^{41}\sin\beta_1 & -\lambda_2 \cdot k_x^{51}\sin\beta_1' - (\lambda_2 + 2\mu_2)k_z^{51}\cos\beta_1' \\ -\mu_2[k_z^{41}\cos\beta_1 - k_x^{41}\sin\beta_1] & -\mu_2[k_z^{51}\sin\beta_1' + k_x^{51}\cos\beta_1'] \end{bmatrix} \tag{7.19}$$

In order to verify the correctness of formula derivation, we now work out the reflected and transmitted coefficients of sliding mass structural surface when SV wave is vertically transmitted onto the surface.

$$A' = \frac{\rho_1 V_{s1} - \rho_2 V_{s2}}{\rho_1 V_{s1} + \rho_2 V_{s2}}; \quad B' = 0; \tag{7.20}$$

$$C' = \frac{2\rho_1 V_{s1}}{\rho_1 V_{s1} + \rho_2 V_{s2}}; \quad D' = 0 \tag{7.21}$$

The calculation results above are exactly the same with the acknowledged calculation results, which demonstrates the derivation results of this chapter are correct.

7.1.4 Instant Seismic Stability Coefficient

Because of the sliding mass structural surface adopting the sliding friction model which conforms to the M-C principle, it should meet the following conditions when coming across shear damage at Point O: φ and C in the formula $\tau_s \geq \sigma_n \tan \varphi + C$ are, respectively, the internal friction angle and cohesive force; if tensile failure occurs at Point O, it should meet the following conditions.

According to the Hooke's law and solving the reflected and transmitted coefficients of seismic wave simultaneously with Formulas (7.17)–(7.19), bring the calculation results and relevant parameters into Formula (7.11), which can get the downslide stress, cohesion stress, and tensile stress of SV wave at Point A. The details are as follows:

$$\begin{aligned}\tau_s = \tau_0 + \mu_1[\mathrm{S}_0^1 \cdot k_z^{11} \cdot \cos\alpha_1 - \mathrm{S}_0^2 \cdot k_z^{21} \cdot \cos\alpha_1 + \mathrm{S}_0^3 \cdot k_z^{31} \cdot \sin\alpha_1' \\ - \mathrm{S}_0^1 \cdot k_x^{11} \cdot \sin\alpha_1 - \mathrm{S}_0^2 \cdot k_x^{21} \cdot \sin\alpha_1 \quad - \mathrm{S}_0^3 \cdot k_x^{31} \cdot \cos\alpha_1']\end{aligned} \tag{7.22}$$

$$\begin{aligned}\sigma_0 \tan\varphi + C = \{\lambda_1[\mathrm{S}_0^1 \cdot k_x^{11} \cdot \cos\alpha_1 - \mathrm{S}_0^2 \cdot k_x^{21} \cdot \cos\alpha_1 + \mathrm{S}_0^3 \cdot k_x^{31} \cdot \sin\alpha_1'] \\ + (\lambda_1 + 2\mu_1)[-\mathrm{S}_0^1 \cdot k_z^{11} \cdot \sin\alpha_1 - \mathrm{S}_0^2 \cdot k_z^{21} \cdot \sin\alpha_1 - \mathrm{S}_0^3 \cdot k_z^{31} \cdot \cos\alpha_1'] + \sigma_0\} \tan\varphi + C\end{aligned} \tag{7.23}$$

$$\begin{aligned}\sigma_n = \lambda_1[\mathrm{S}_0^1 \cdot k_x^{11} \cdot \cos\alpha_1 - \mathrm{S}_0^2 \cdot k_x^{21} \cdot \cos\alpha_1 + \mathrm{S}_0^3 \cdot k_x^{31} \cdot \sin\alpha_1'] + (\lambda_1 + 2\mu_1) \\ [-\mathrm{S}_0^1 \cdot k_z^{11} \cdot \sin\alpha_1 - \mathrm{S}_0^2 \cdot k_z^{21} \cdot \sin\alpha_1 - \mathrm{S}_0^3 \cdot k_z^{31} \cdot \cos\alpha_1'] + \sigma_0\end{aligned} \tag{7.24}$$

Because the sliding mass structural surface conforms to the M-C principle, it should meet the following conditions when coming across shear damage at Point A: φ and C in the formula are, respectively, the internal friction angle and cohesive force; if tensile failure occurs at Point A, it should meet the following conditions: σ_{tension} in formula $\sigma_n \geq \sigma_{\text{tension}}$ stands for the permitted tensile stress. In Formula (7.22) to (7.24), σ_0 and τ_0 are, respectively, the initial normal and tangential gravitational stress at Point A, so it can be solved with slice method. By analogy, the above method could solve the stress of other points on the sliding mass structural surface. Then, collect the calculation results of each point to solve the total downslide stress SF and the total tensile stress TF on the sliding mass structural surface [see Formula (7.25)]. Finally, use the safety coefficient KO required by the design to judge the slope safety [see Formula (7.26)].

$$F_1 = \sum_{i=1}^{n} \sigma_{0i} \cdot \tan\varphi_i + C_i \cdot \mathrm{d}A_i \tag{7.25}$$

$$\begin{aligned}&\text{If } K = \mathrm{TF}/\mathrm{SF} < K_0, \text{the sliding mass is dangerous;} \\ &\text{if } K = \mathrm{TF}/\mathrm{SF} > K_0, \text{the sliding mass is safe.}\end{aligned} \tag{7.26}$$

In Formulas (7.25) and (7.26), τ_{si} stands for the downslide shear stress of the ith meshing point; $A_i = 0.5\ (L_{i-1} + L_{i+1}) \times 1$, which represents the area of the ith meshing point, in which L_{i-1} and L_{i+1} are, respectively, the side length of the $i - 1$th and $i + 1$th meshing points; σ_{0i} represents the normal stress of the ith meshing point; and φ_i and C_i, respectively, stand for the internal friction angle and cohesive force of the ith meshing point.

Through comprehensive analysis of Formula (7.22) to (7.26), the seismic stability time–frequency analysis method derived from this chapter could rationally consider (1) gravitational stress field, (2) the internal friction angle, cohesive force, and reflected and transmitted coefficients of controlling structural surface, (3) physico-mechanical parameters of bedrock and cover layer, and (4) influencing factors such as instant frequency, incident angle, acceleration, velocity, and displacement–time history curve.

7.1.5 Time–Frequency Analysis Process of Seismic Waves

Through comprehensive analysis of the seismic stability time–frequency analysis method of bedrock and overburden slopes, it can be known that the instant frequency effects of seismic waves mainly embody the elastic displacement amplitude s_0^i of incident wave, reflected wave, and transmitted wave and the frequency of wave vector ω, i.e., k_x^{i1} and k_z^{i1}, for example $k_x^{i1} = \omega \cdot \sin\alpha_1 / V_{p2}$. Therefore, this dissertation proposes the analysis process of seismic wave time–frequency effects: firstly, break up seismic wave time–frequency into several of IMF (time history curve) through empirical mode decomposition (EMD for short); secondly, carry out channel switch, i.e., switch multi-channel signal composed of multiple IMF into single-channel signal made up of single IMF; thirdly, solve the instant frequency of each channel, and draw the time–frequency curve of each IMF. Lastly, bring each IMF and their instant frequency into Formulas (7.22)–(7.26) to work out the results and superimpose the results, and we can get the stress component of this point.

7.1.6 Practical Application of Seismic Wave Time–Frequency Effects

In order to more deeply and concretely explain the practical application of seismic wave time–frequency effects, this dissertation will, based on the above analysis

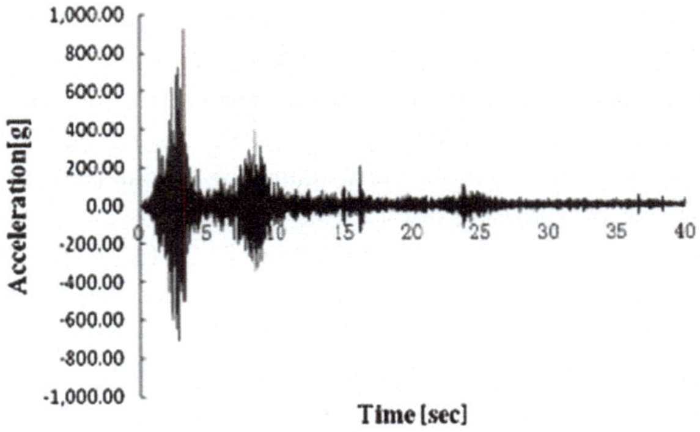

Fig. 7.3 The time history curve of input seismic wave

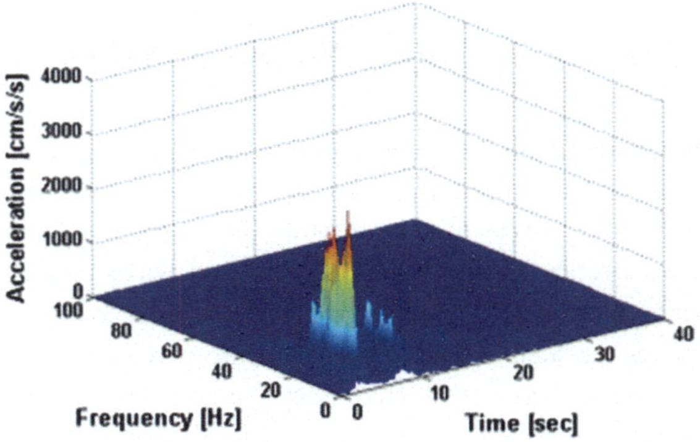

Fig. 7.4 Hilbert spectrum of horizontal acceleration

ideas, conduct input wave time–frequency analysis and the detail is as follows: Because the follow-up work needs combining the shaking table test and numerical analysis model to verify the correctness of this time–frequency method, this chapter will select the original horizontal seismic waves of the shaking table test, i.e., horizontal Wenchuan wolong seismic wave (see Fig. 7.3), in which the Hilbert spectrum is shown Fig. 7.4; secondly, carry out EMD decomposition and get every IMF (see Fig. 7.5); and lastly, solve the instant frequency of every IMF (see Fig. 7.6).

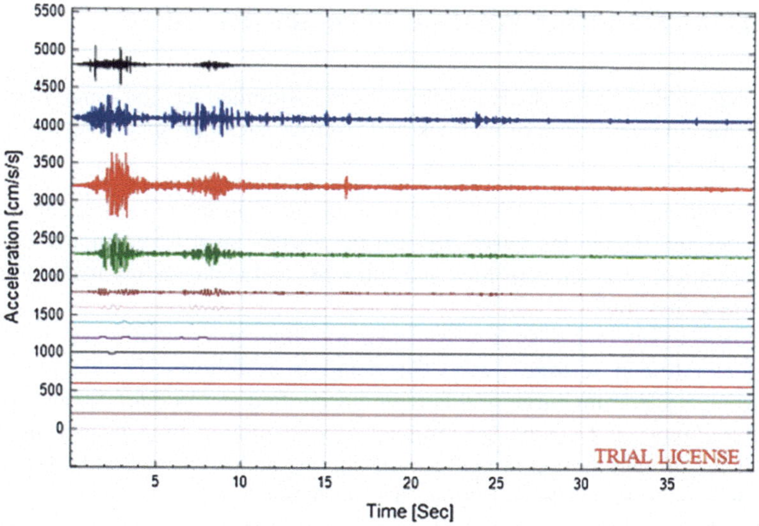

Fig. 7.5 Acceleration time history curve of IMF. *Note* The *Y* coordinate only represents physical characters, but not size

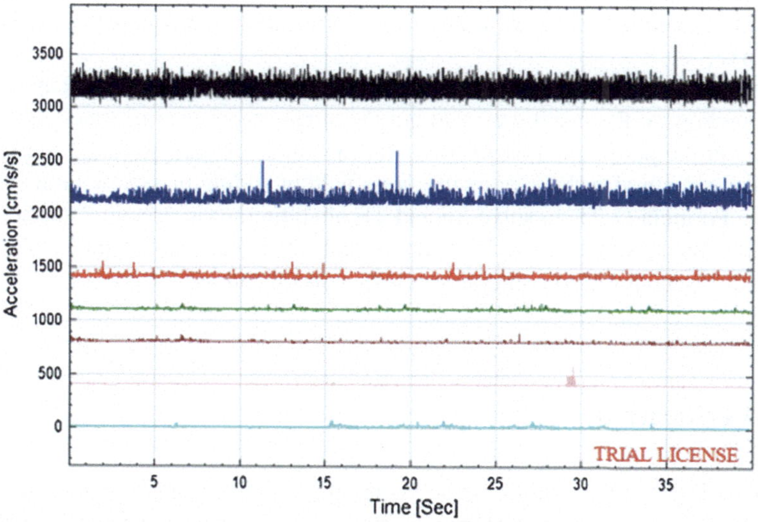

Fig. 7.6 Instant frequency map of IMF. *Note* The *Y* coordinate only represents physical characters, but not size

7.2 Time–Frequency Analysis Formula Derivation

According to the above analysis, this chapter summarizes the time–frequency analysis formula derivation of bedrock and overburden slope seismic stability, and it is as follows:

(1) According to the study object, we sum up the appropriate geological analysis model and decide the basic physical parameters such as geometric size and mechanical parameters of rock and soil mass and make sliding mass structural surface meshing;
(2) If there exists real seismic wave, then we adopt it as input wave; if not, we adopt artificial seismic wave. The detail is as follows: Decide the design response spectrum curve of this region according to *Code for Seismic Design of Buildings* and synthesize artificial ground motion time history based on seismic ground motion time history fitting method, so as to decide the seismic ground motion acceleration time history of this region.
(3) Decide the displacement–time history curve according to the seismic ground motion time history curve, adopt any points in the mesh as study point, and use Formula (7.17) to solve the seismic ground motion time history curve of reflected wave and transmitted wave at this point.
(4) Use Hilbert–Huang transform solving IMF of incident wave, reflected wave, and transmitted wave as well as the corresponding frequency–time history curve;
(5) Bring the obtained results into Formulas (7.22)–(7.24) to calculate and superimpose the results, we can get the stress component of this point;
(6) Adopt the above method to work out the stress of each point on the sliding mass structural surface, draw the stress field of sliding mass structural surface, use Formula (7.25) to solve SF and TF of sliding mass structural surface, and judge the seismic stability of bedrock and overburden slopes under SV wave with Formula (7.26) combining relevant specifications.

7.3 Verification

Due to the fact that the monitoring points of shaking table test could not directly be set up on the sliding mass structural surface, which cannot accurately judge the damage development of the sliding mass structural surface, it can only reflect the whole damage of the slope and indirectly reflect the damage of the sliding mass structural surface. Therefore, this chapter will combine the shaking table test with the numerical analysis method and verify the correctness of the above time–frequency method from two perspectives of sliding mass and structural surface. Before verifying the correctness of seismic stability time–frequency analysis method of bedrock and overburden slopes, this chapter needs to explicitly explain the verification approach, and the details are as follows: First, build numerical analysis model by geometric similarity ratio of 1:1 to the shaking table test model of double-sided bedrock and overburden slopes, set up stress and strain test points at the corresponding position, and compare the numerical calculation results with the experiment results so as to verify the correctness of numerical analysis method. This part of work has been finished in Sect. 4.3. Secondly, carry out explicit and complete verification of the calculation results of time–frequency analysis method using the shaking table test in Chap. 4 and the above numerical analysis results.

7.3.1 Overview

In order to accurately verify the correctness of seismic stability time–frequency analysis method of bedrock and overburden slopes, this chapter will build the numerical analysis method according to the shaking table test and adopt same rock and soil parameters and input seismic wave (see Figs. 4.1, 4.9, and 4.10). In addition, the bedrock and cover layer adopt the elastoplastic model and the sliding mass structural surface adopts the brittle fracture constitutive model, and both models conform to the M-C failure principle. Set up stress and strain Test Point A on the sliding mass structural surface. And at the same time, the seismic stability time–frequency analysis method of bedrock and overburden slopes also select Test Points A and B as study objects, so as to accurately and completely verify the correctness of calculation results of single point on the structural surface and the sliding mass. See Fig. 7.7 for test point distribution. It is noteworthy that because the Wenchuan wolong seismic wave mainly includes P wave and S wave, S wave includes SV wave and SH wave, in which the seismic wave energy carried by P

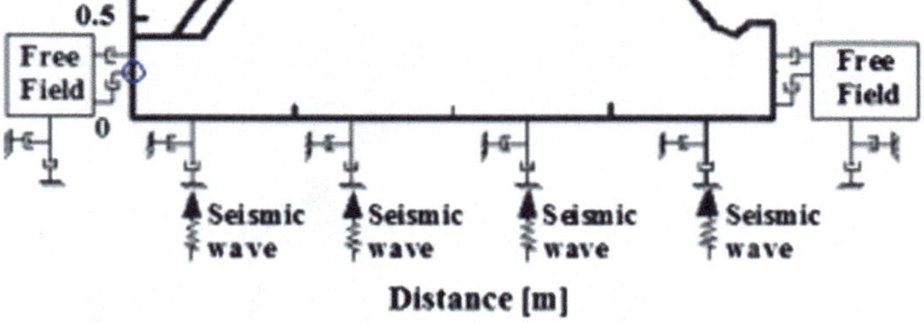

Fig. 7.7 Numerical analysis model

wave is small and that carried by A wave is big, which plays a key role in the slope seismic stability. At the same time, P wave and SV wave mainly cause damage inside the cross section, while SH wave mainly causes damage outside the cross section. However, the slope problem is often simplified into plane stress problem, which pays no attention to the slope damage outside the cross section, so it does not need to consider the effects of SH wave. In general situation, most researches of slope seismic stability only consider the effects of SV wave and approximately take the real seismic wave as SV wave to partially consider the safety. Therefore, this shaking model test approximately simulates the effects of SV wave through input of Wenchuan wolong seismic wave.

7.3.2 Gravitational Stress Field

This chapter regulates that the normal stress takes stress as positive and tension as negative; shear stress takes the upward of sliding mass structural surface as positive and upward as negative.

Test Point A is 20 cm from the surface of cover layer, the tangential obliquity of sliding mass structural surface at micro-unit A is 32°, and this chapter adopts slice method to calculate the gravitational stress field (see Figs. 7.8, 7.9 and 7.10). Lastly, we get the initial normal stress σ and initial shear stress τ of this point as, respectively, 3.29 and 1.90 kPa, so according to the M-C failure principle, it can be known that the rest sliding shear stress P is as follows:

$$P = \sigma \tan \theta + C - \tau = 3.29\,\text{kPa} \times \tan 32^\circ + 2.3\,\text{kPa} - 1.9\,\text{kPa} = 2.455\,\text{kPa}$$

7.3.3 Calculation Results' Comparison

Based on the derivation process of seismic stability time–frequency analysis method of bedrock and overburden slopes, we bring the relevant parameters and the

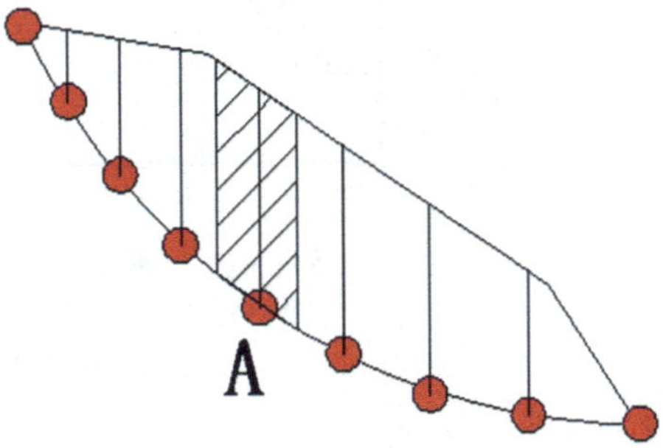

Fig. 7.8 Analysis model of deposit body

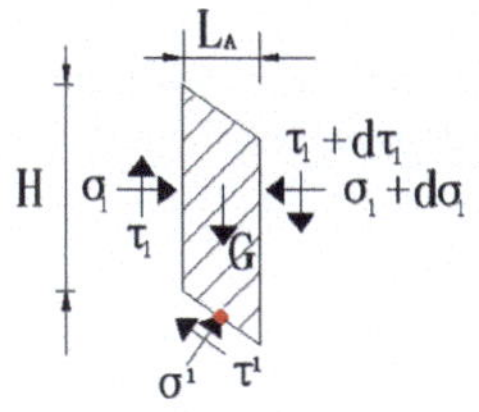

Fig. 7.9 Mechanical model figure of sliding block A

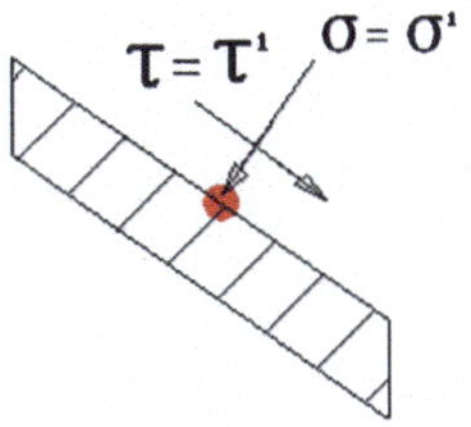

Fig. 7.10 Mechanical model figure of Test Point A

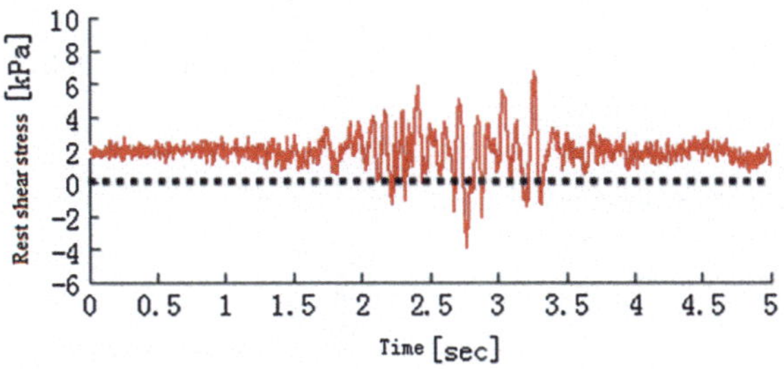

Fig. 7.11 Rest shear stress time history curve of Test Point A

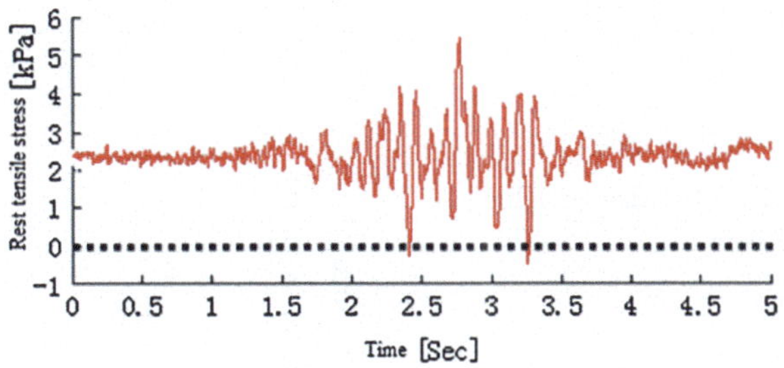

Fig. 7.12 Rest tensile stress time history curve of Test Point A

above calculation results into Formula (7.22) to (7.24) to solve the stress of Test Point A, so as to work out the rest sliding stress of it (see Figs. 7.11 and 7.12 for detailed results). At the same time, for the calculation results of plastic stress of Test Point A, see Fig. 7.13 and work out the rest sliding stress of left points by analogy. Finally, we superimpose the results and get the rest sliding stress–time history curve of the whole sliding mass structural surface, and the detailed calculation results are shown in Figs. 7.14 and 7.15. At the same time, the displacement calculation results of test Point are shown in Fig. 7.16. Due to the fact that the duration of Wenchuan

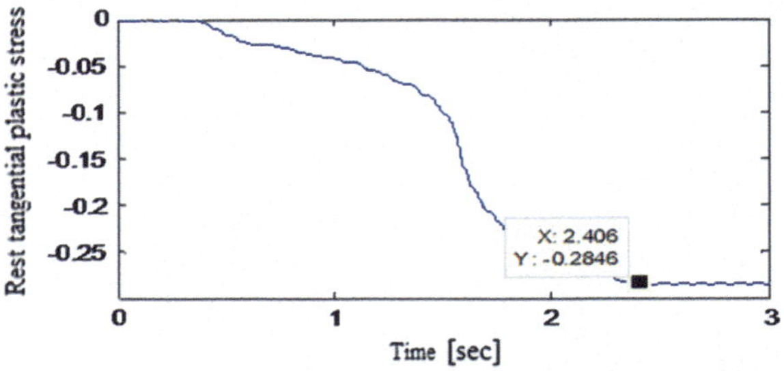

Fig. 7.13 Rest tangential plastic stress–time history curve of Test Point A

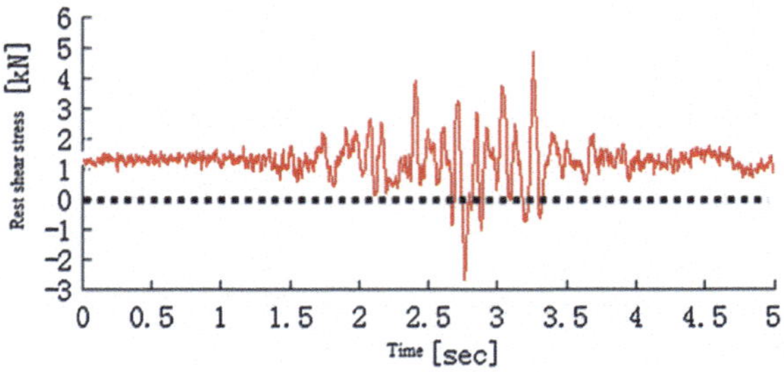

Fig. 7.14 Rest shear stress–time history curve of the sliding mass

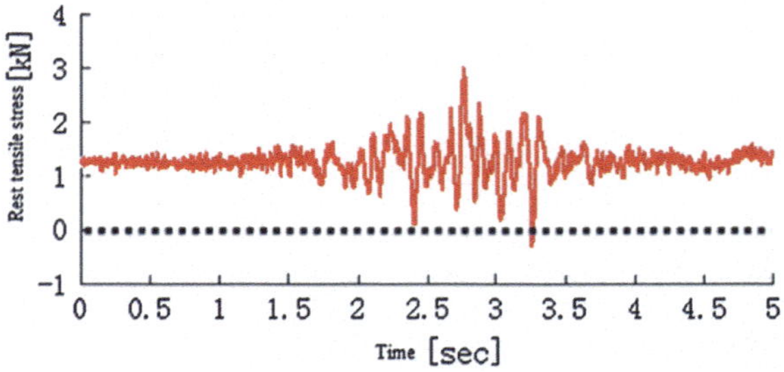

Fig. 7.15 Rest tensile stress–time history curve of the sliding mass

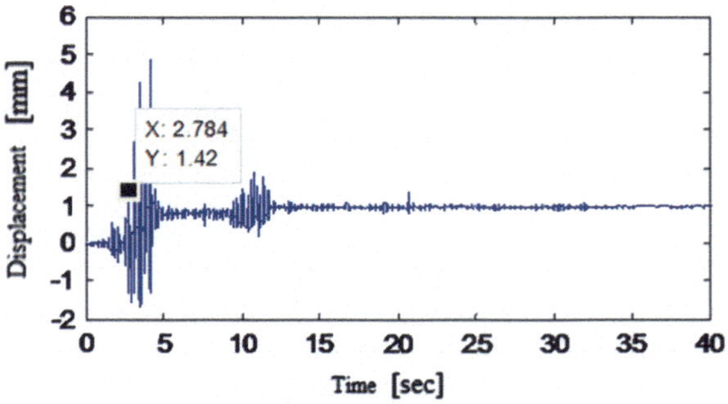

Fig. 7.16 Shaking table test results

wolong seismic wave is a bit longer and Test Point A is damaged before $T = 5$ s, Figs. 7.11, 7.12, 7.13, 7.14, 7.15, and 7.16 only select calculation results of $T =$ 0–5 s including peak seismic ground motion acceleration.

It can be known from Figs. 7.11 and 7.12 that Test Point A comes across shear failure at $T = 2.25$ s and tensile failure at $T = 2.45$ s. Therefore, Test Point A will come across shear failure at $T = 2.25$ s in the earthquake; the numerical analysis results in Fig. 7.13 show that when $T = 1$–2 s, the tangential plastic stress of A suddenly changes and becomes stable after $T = 2.406$ s. According to the features of the brittle fracture constitutive model and M-C failure principle, we can know that the test point has already been damaged. It can be known from Figs. 7.14 and 7.15 that the sliding mass coming across shear failure at $T = 2.75$ s and tensile failure at $T = 3.25$ s, and thus, in the earthquake, the sliding mass will come across shear failure at $T = 2.75$ s; the results of the shaking table test in Fig. 7.16 show that the displacement of the test point suddenly changes, which can demonstrate that the sliding mass has already been damaged. The difference of the calculation results with the numerical analysis results and the shaking table test results may be because firstly, the time–frequency analysis method of rock slope seismic stability was derived based on the elastic wave theory, which paid no attention to the nonlinear characteristics of rock and soil mass as well as the sliding mass structural surface; secondly, the time–frequency analysis method of rock slope seismic stability did not consider the plastic strain and accumulative effect in the earthquake. But in general, no matter the single test point or the whole sliding mass, the calculation results of time–frequency analysis method, the numerical analysis results, and the shaking table test results of rock slope seismic stability are basically the same, which fully verify the correctness of the calculation results and feasibility of the calculation method in this dissertation.

Through comprehensive analysis of the above calculation results, it can be known that under SV wave effects, the sliding mass structural surface mainly comes across shear failure, which is basically the same with the results of field research of Wenchuan earthquake; the time of sliding mass structural surface coming across failure is slightly ahead of the PGA arriving time of $T = 3.25$ s in Wenchuan earthquake, which fully explains that it is a little bit conservative for the current supporting structure seismic design adopting the method of PGA, so this chapter suggests using time–frequency design so as to save and reduce cost.

7.4 Study on Parameters

Through comprehensive analysis of the above derivation, it can be known that the transmission and reflection coefficients of structural surface include factors like the elastic displacement amplitude of incident wave, reflected wave, and transmitted wave. Therefore, it is quite necessary to conduct research on the influence factors such as transmitted and reflected coefficients of the structural surface in this chapter.

7.4.1 Incident Angle on Transmitted and Reflected Coefficients of the Structural Surface

In order to study the influence of incident angle on transmitted and reflected coefficients of the structural surface, this chapter now takes the following example to explain: Take sine wave with amplitude = 1 and frequency = 5 Hz as the incident S wave, and the incident angles are 10°, 20°, 30°, 40°, 50°, 60°, 70°, 80°, and 90°. When the elasticity modulus of the cover layer, respectively, is 20, 100, 500, and 1000 MPa, the elasticity modulus of the bedrock is 30,000 MPa. The reflected and transmitted coefficients of different structural surface parameters are shown in Figs. 7.17, 7.18, 7.19, and 7.20.

Through comprehensive analysis of Figs. 7.17, 7.18, 7.19, and 7.20, we can get the following rules: The reflected coefficient of reflected S wave reaches the maximum when the incident angle is 38°, and the phase position changes when the incident angle is 25°; the reflected coefficient of reflected P wave slightly grows with the increase in the incident angle when the incident angle is in the range of 0°–38°, while it suddenly increases when the incident angle is 38°, and when the incident angle is in the range of 0°–40°, the phase position of the reflected coefficient stays unchanged; the transmitted coefficient of transmitted S wave slightly decreases with the increase in the incident angle when the incident angle is in the range of 0°–38° and reaches the maximum when the incident angle is 38°, while it suddenly increases when the incident angle is in the range of 38°–40°, and when the incident angle is in the range of 0°–40°, the phase position of the transmitted coefficient stays unchanged; the transmitted coefficient of transmitted P wave reaches its maximum when the incident angle is 38° and suddenly increases afterward.

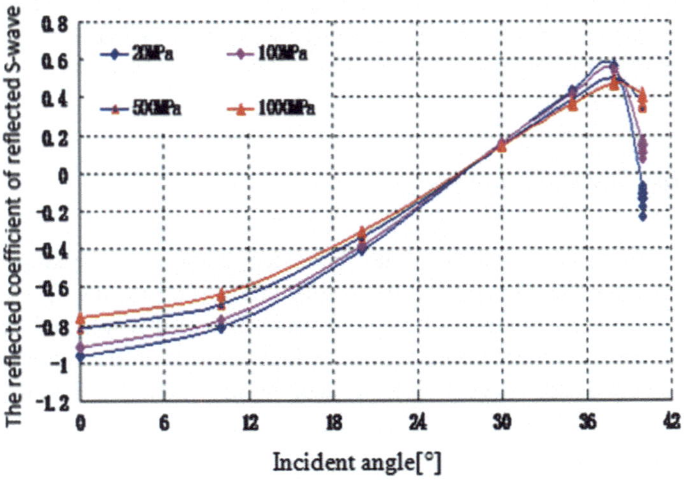

Fig. 7.17 The reflected coefficients of reflected S wave with different incident angles

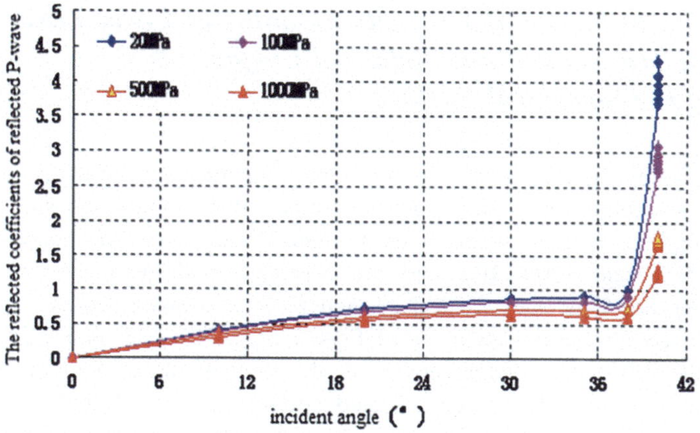

Fig. 7.18 The reflected coefficients of reflected P wave with different incidents

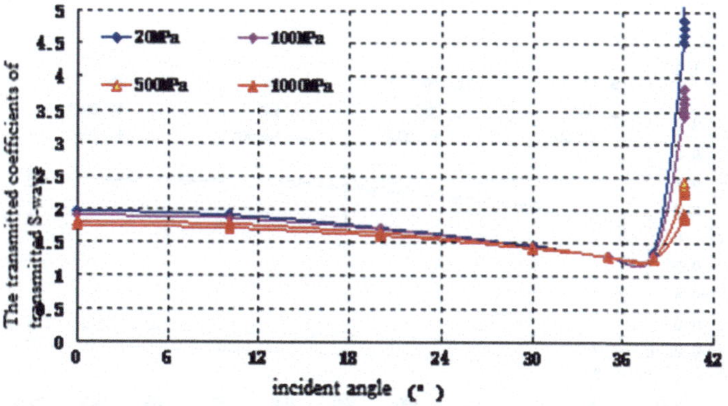

Fig. 7.19 The transmitted coefficients of transmitted S wave with different incident angles

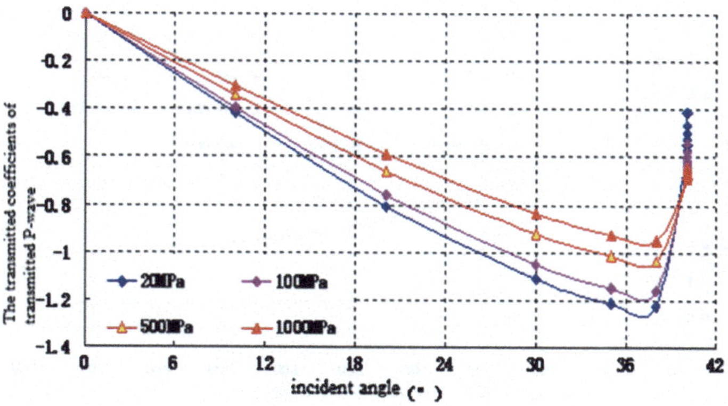

Fig. 7.20 The transmitted coefficients of transmitted P wave with different incident angles

7.4.2 *The Influence of Elasticity Modulus of Cover Layer on Transmitted and Reflected Coefficients of the Structural Surface*

In order to study the influence of elasticity modulus of cover layer on transmitted and reflected coefficients of the structural surface, this chapter now still takes the above example to explain, the elasticity modulus of the cover layer, respectively, is 20, 100, 500, and 1000 MPa, and the elasticity modulus of the bedrock is 30,000 MPa. See Figs. 7.21, 7.22, 7.23, and 7.24 for detailed results.

Through comprehensive analysis of Figs. 7.21, 7.22, 7.23, and 7.24, we can come to the following conclusions: With the gradual increase in the elasticity modulus of the cover layer, the reflected coefficients of reflected S wave, the reflected coefficients of reflected P wave, the transmitted coefficients of transmitted S wave, and the transmitted coefficients of transmitted P wave all come to certain balance, which are approximately in hyperbolic curve or inverted hyperbolic curve.

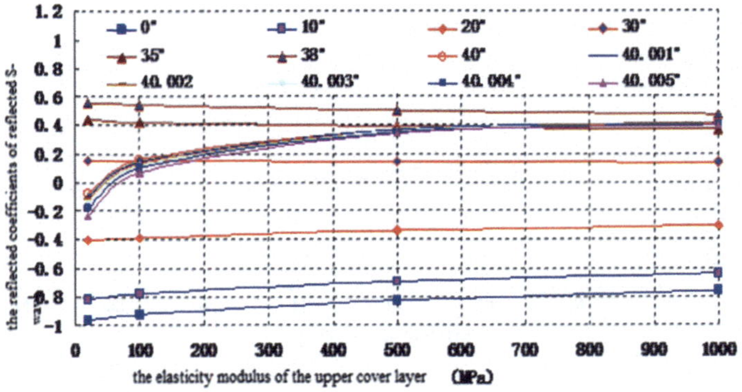

Fig. 7.21 The reflected coefficients of reflected S wave

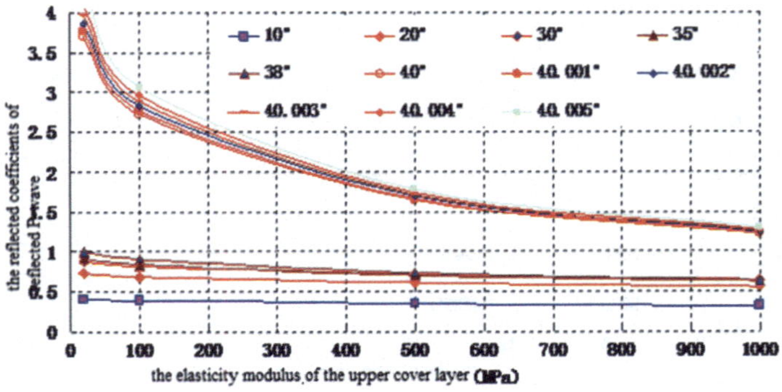

Fig. 7.22 The reflected coefficients of reflected P wave

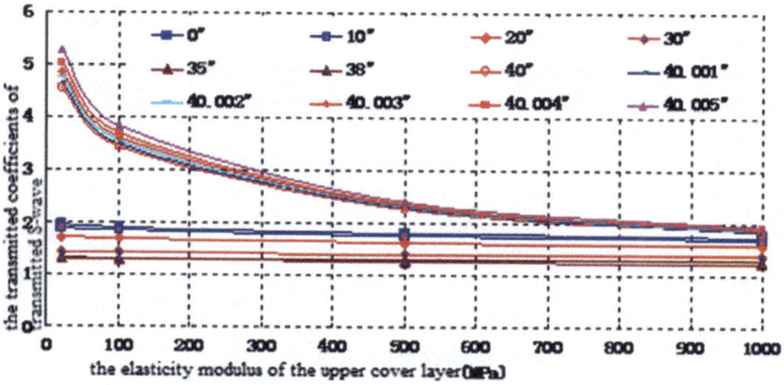

Fig. 7.23 The transmitted coefficients of transmitted S wave

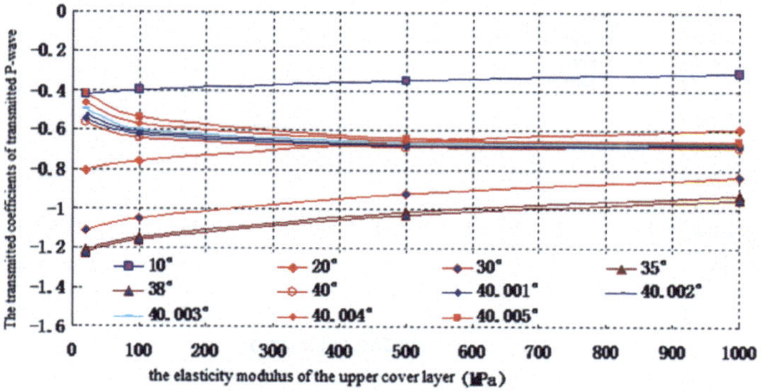

Fig. 7.24 The transmitted coefficients of transmitted P wave

7.4.3 The Influence of Incident Angle on Transmitted and Reflected Coefficients of the Structural Surface

In order to study the influence of incident angle on transmitted and reflected coefficients of the structural surface, this chapter now takes the above example to explain and the incident angles are 10°, 20°, 30°, 40°, 50°, 60°, 70°, 80°, and 90°. See Figs. 7.25, 7.26, 7.27, and 7.28 for detailed results.

Through comprehensive analysis of Figs. 7.25, 7.26, 7.27, and 7.28, we can come to the following conclusions: The energy reflection coefficient of reflected S wave reaches its maximum when the incident angle is 0°, and arrives at a small peak when the incident angle is 38°, but it is close to 0 when the incident angle is 30°. The energy reflection coefficient of reflected P wave reaches its maximum when the incident angle is 30°. The energy transmission coefficient of transmitted S wave reaches its maximum when the incident angle is 0° and lightly decreases when the incident angle is in the range of 0°–38°, but when the incident angle is in the range of 38°–40°, it suddenly increases. The energy transmission coefficient of

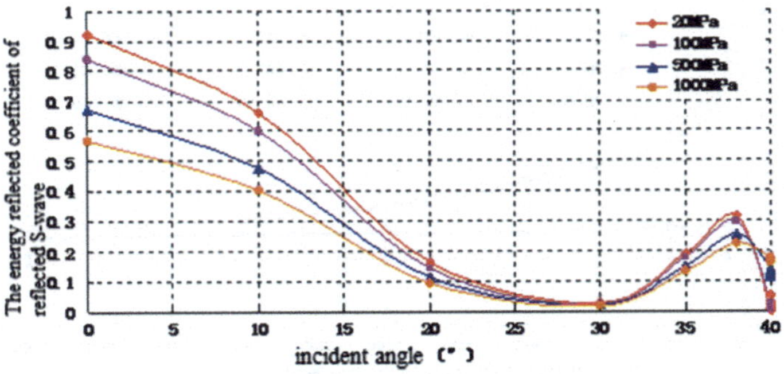

Fig. 7.25 The energy reflection coefficient of reflected S wave

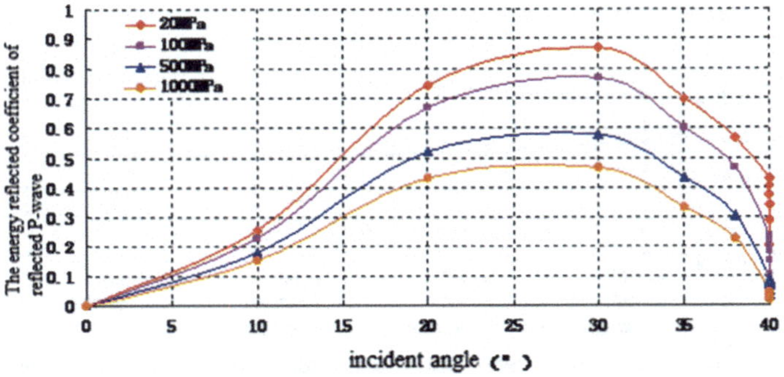

Fig. 7.26 The energy reflection coefficient of reflected P wave

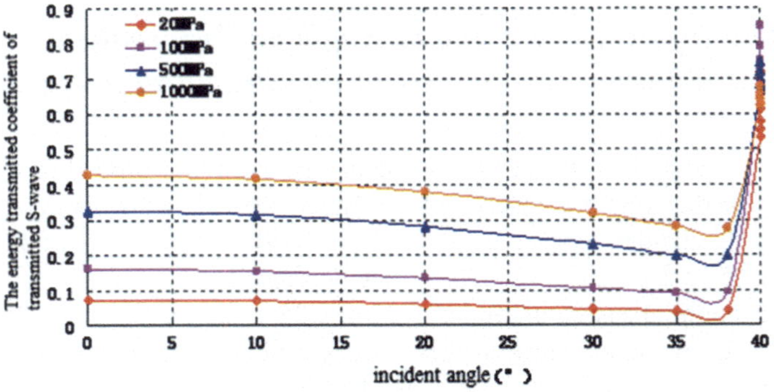

Fig. 7.27 The energy transmission coefficient of transmitted S wave

transmitted P wave reaches its maximum when the incident angle is 38°, it increases with the increase in incident angle when the latter is in the range of 0°–38°, and it decreases with the increase in incident angle when the latter is in the range of 38°–40°.

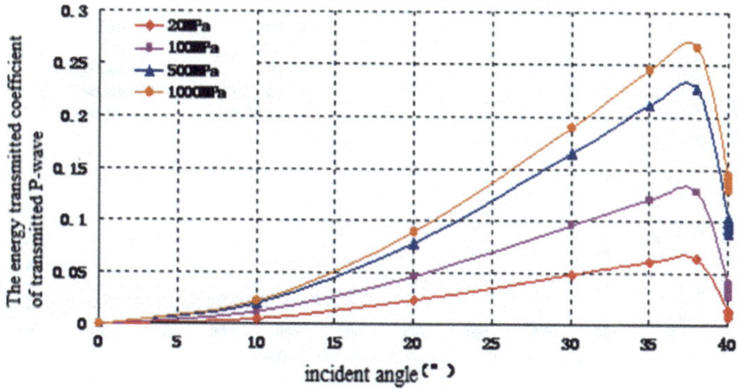

Fig. 7.28 The energy transmission coefficient of transmitted P wave

7.4.4 The Influence of Elasticity Modulus of Cover Layer on Transmitted and Reflected Coefficients of the Structural Surface

In order to study the influence of elasticity modulus of cover layer on transmitted and reflected coefficients of the structural surface, this chapter now still takes the above example to explain, the elasticity modulus of the cover layer, respectively, is 20, 100, 500, and 1000 MPa, and the elasticity modulus of the bedrock is 30,000 MPa. See Figs. 7.29, 7.30, 7.31, and 7.32 for detailed results.

Through comprehensive analysis of Figs. 7.29, 7.30, 7.31, and 7.32, we can come to the following conclusions: The energy reflection coefficient of reflected S wave and the energy reflection coefficient of reflected P wave all decrease with the increase in elasticity modulus of the cover layer and finally reach the balance; the energy transmission coefficient of transmitted S wave increases with the increase in elasticity modulus of the cover layer and finally becomes stable. The energy

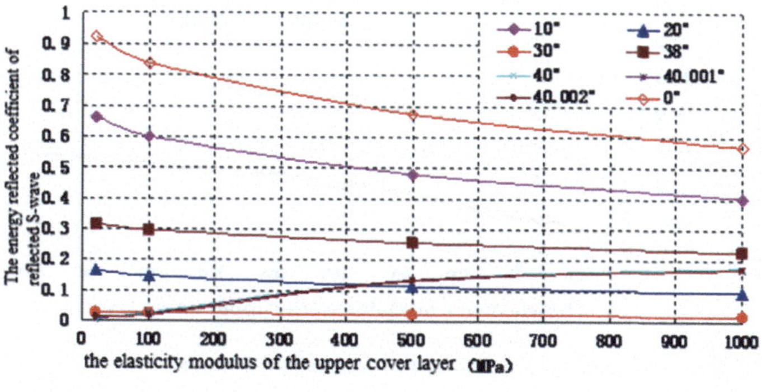

Fig. 7.29 The energy reflection coefficient of reflected S wave

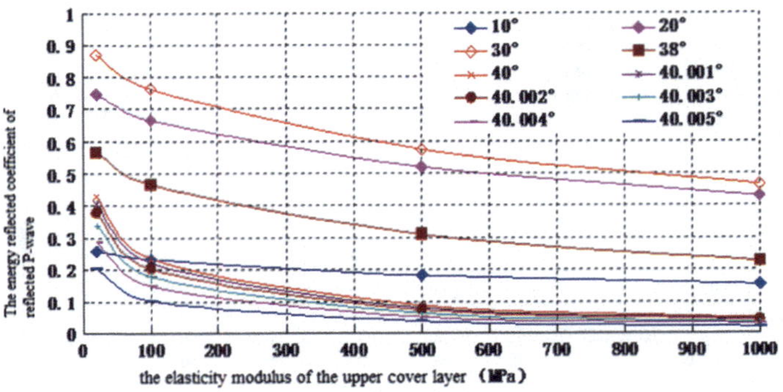

Fig. 7.30 The energy reflection coefficient of reflected P wave

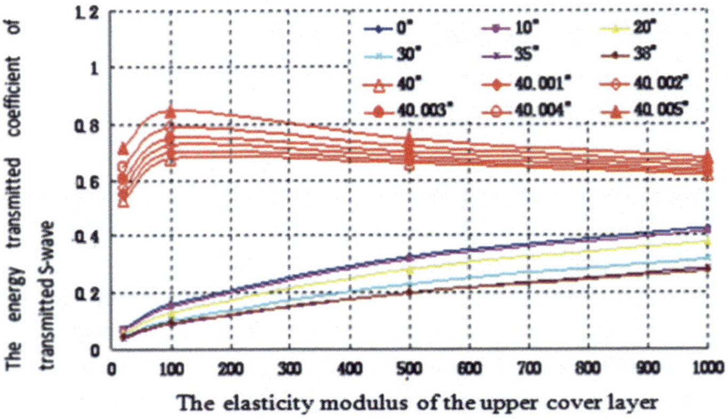

Fig. 7.31 The energy transmission coefficient of transmitted S wave

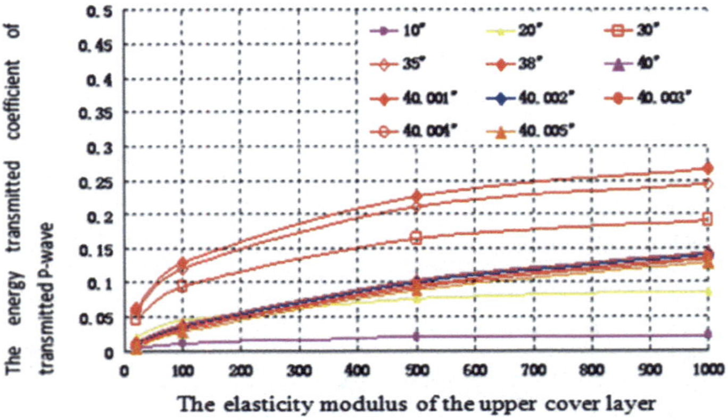

Fig. 7.32 The energy transmission coefficient of transmitted P wave

transmission coefficient of transmitted P wave reaches its maximum when the thickness of cover layer reaches 2000 MPa and slightly decreases afterward.

7.4.5 The Influence of Elasticity Modulus of Cover Layer on Total Transmitted and Reflected Energy Coefficients of the Structural Surface

In order to study the influence of elasticity modulus of cover layer on total transmitted and reflected energy coefficients of the structural surface, this chapter now still takes the above example to explain, the elasticity modulus of the cover layer, respectively, is 20, 100, 500, and 1000 MPa, and the elasticity modulus of the bedrock is 30,000 MPa. See Figs. 7.33 and 7.34 for detailed results.

Through comprehensive analysis of Figs. 7.33 and 7.34, it can be known that the total reflected energy coefficients slightly decrease with the increase in incident angle when the latter is in the range of 0°–40° and reach its minimum when the

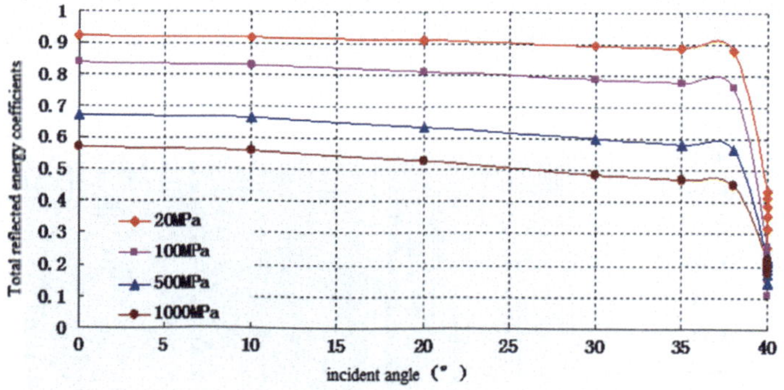

Fig. 7.33 Total reflected energy coefficients

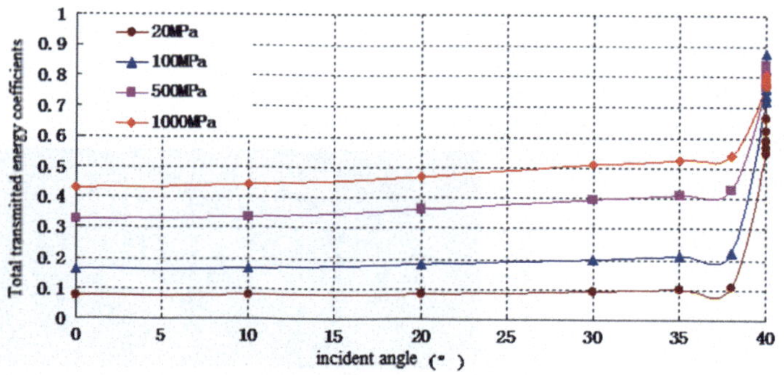

Fig. 7.34 Total transmitted energy coefficients

incident angle is 38°. Then, with the continuous increase in incident angle, the total reflected energy coefficients suddenly increase; the total reflected energy coefficients are precisely opposite to the total energy reflection, and with any incident angles, the sum of them is 1.

7.5 Advantage

In order to fully explain the advantage of seismic stability time–frequency analysis of bedrock and overburden slopes, this chapter selects sine wave whose frequency changes with time (see Figs. 7.35 and 7.36 for detailed results) and uses time–frequency analysis method, pseudo-static method, and numerical analysis method to calculate the slope stability (see Fig. 7.37 for analysis model). The concrete physical and mechanical indexes of bedrock, upper cover layer, and the sliding mass structural surface are given in Table 7.1. It is noteworthy that because the sliding block method could not concretely and quantitatively judge the seismic stability of slopes, this chapter will not take comparison of it.

Based on the solving process of seismic stability time–frequency analysis of bedrock and overburden slopes, we bring the relevant parameters and the above calculation results into Formula (7.22)–(7.24) to work out the stress of Point A, so

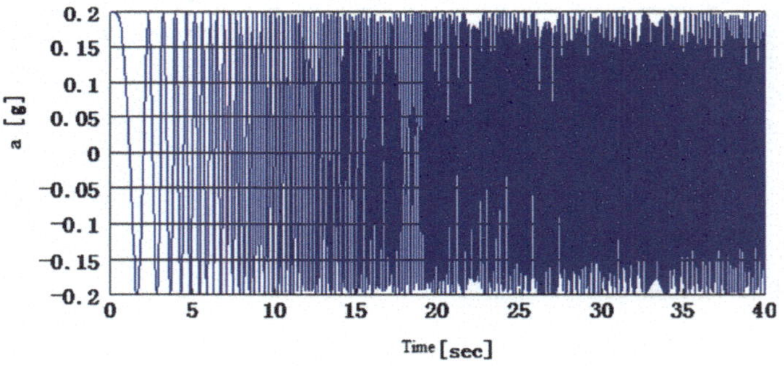

Fig. 7.35 Time history curve of sine wave

Fig. 7.36 Hilbert response spectrum of sine wave

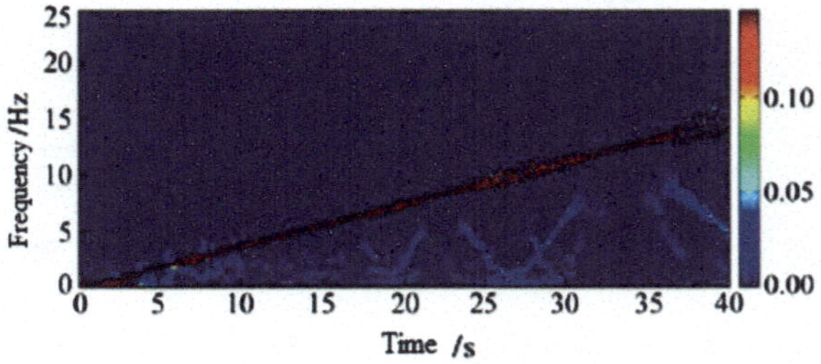

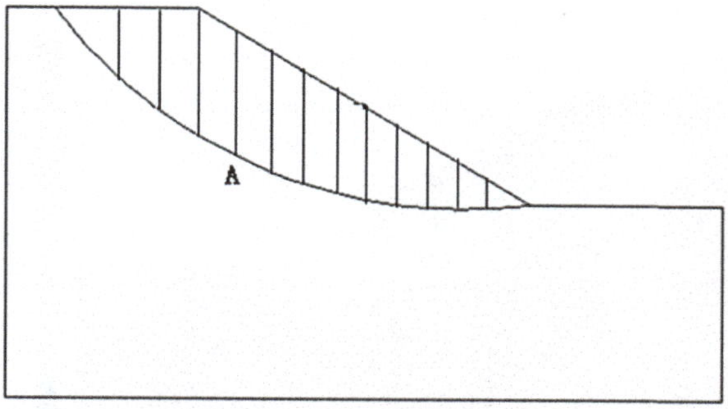

Fig. 7.37 Slope analysis model

Table 7.1 The physico-mechanical parameters of bedrock, upper cover layer, and the sliding mass structural surface

Model name	Unit weight	Shear wave speed	P wave speed	Lame coefficient	Shear modulus
Bedrock	30 kN/m^3	826.19 m/s	3240 m/s	881.89 MPa	2047.24 MPa
Upper cover layer	19 kN/m^3	99.21 m/s	145.34 m/s	6.30 MPa	6.30 MPa
The structural surface	Normal stiffness	Shear stiffness	Internal friction angle	Tensile strength	Cohesive force
	4500 MPa	2300 MPa	32°	23 kPa	50

as to calculate the rest sliding shear stress of it. Calculate the rest sliding shear stress of the left points by analogy and finally superimpose the results, and we can get the shear sling stress–time history curve of the whole sliding mass structural surface (see Fig. 7.38 for detailed results). See Fig. 7.39 for the calculation results of numerical analysis method. At the same time, we select PGA = 0.2 g and calculate the rest shear sliding stress with pseudo-static method, and we can get the result as −47 kN, so the slope is safe.

It can be known from Fig. 7.38 that when T = 10.0 s, the rest shear sliding stress of calculation results with time–frequency analysis method is negative and the slope

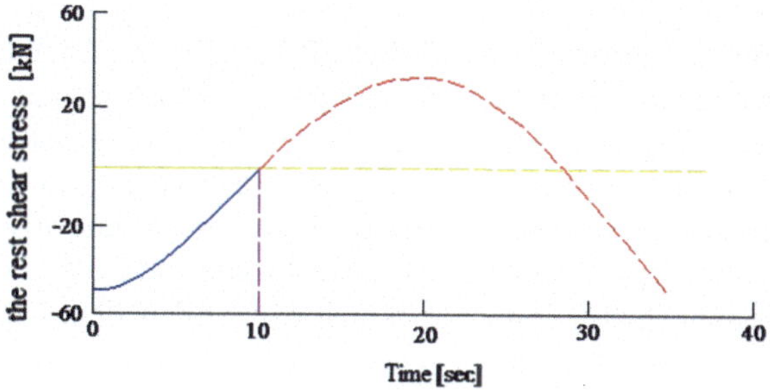

Fig. 7.38 Calculation results of time–frequency analysis method

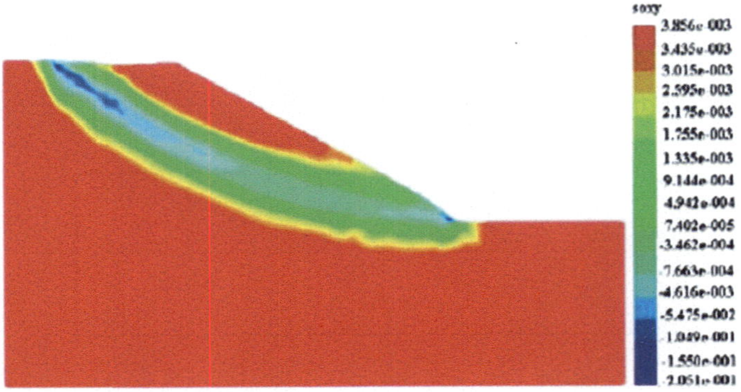

Fig. 7.39 Numerical calculation results (shear plastic strain figure)

has been damaged. While due to that the time–frequency analysis method in this chapter is derived through elastic wave theory which pay no attention to the non-linear characteristics of rock and soil mass, so there still has calculation results when $T = 10.0–35.0$ s, but these results are invalid and can only display the rest shear sliding stress under different seismic wave frequency. Figure 7.39 shows that when the numerical calculation results display at $T = 9.3$ s, shear plastic zone occurs inside the slope, and the slope is damaged. Therefore, the results of time–frequency analysis method and numerical analysis method are basically the same, while the calculation results of pseudo-static method show that the slope is safe and will not be damaged, which is opposite to the results of time–frequency analysis method and numerical analysis method. The above phenomenon may be caused by the following reasons: First, the natural frequency of the sliding mass is obtained by the related formula proposed by Xu Jiyan and Liu Hongshuai et al. With the duration of input seismic wave, its frequency gradually increases and approaches the natural frequency of the sliding mass. The resonance effects become stronger, while the rest shear sliding stress decreases. When $T = 10.0$ s, the instant frequency of seismic wave is 4.9 Hz, the rest shear sliding stress decreases to negative, and the slope slides and failure occur; second, for time–frequency analysis method and numerical analysis method, they can not only consider the influence of PGA on slope stability, but also consider the influence of frequency on the slope stability, while for pseudo-static method, it can only consider the influence of PGA on slope stability, which leads to the difference among the above calculation results; lastly, time–frequency analysis method can pay no attention to the plastic strain and accumulative effect in the earthquake, which caused the result displacing time of time–frequency analysis method lagging behind that of numerical analysis method.

Based on the above analysis, it can be known that the frequency of seismic ground motion has significant influence on the seismic stability of bedrock and overburden slopes, so we need to take into consideration when calculating the seismic stability of bedrock and overburden slopes.

7.6 Brief Summary

This chapter derives the seismic stability time–frequency analysis method of bedrock and overburden slope and verifies its rationality through shaking table test and numerical analysis results. This method could not only consider the influence of three essential elements (PGA, frequency, and duration) of SV wave on slope stability, but also predict the safety of rock slopes, the happening time, and scale of landslides. At the same time, it can provide reference to support structure seismic time–frequency design in high-intensive seismic regions, and we come to the following conclusions:

(1) Adopting the seismic stability time–frequency analysis method of bedrock and overburden slope to predict the damage of the shaking table test model of one rock slope on Zipingpu Reservoir bank, it rationally predicts the happening time of landslides, which is basically the same with the results of shaking table test and numerical analysis results and fully verifies the feasibility of this method. At the same time, this method could not only properly predict the seismic landslides, but also provide valuable reference to support structure seismic time–frequency design in high-intensive seismic regions.
(2) Through carrying out comparative analysis of time–frequency analysis method, pseudo-static method, and numerical analysis method, the result reveals that time–frequency analysis method could consider the influence of seismic time history and frequency on slope stability.
(3) The seismic stability time–frequency analysis method of bedrock and overburden slope has certain universality and comprehensiveness, and it could rationally consider (1) gravitational stress field, (2) the internal friction angle, cohesive force, and reflected and transmitted coefficients of controlling structural surface, (3) physico-mechanical parameters of bedrock and cover layer, and (4) influencing factors such as instant frequency, incident angle, acceleration, velocity, and displacement–time history curve.
(4) The elasticity modulus of incident angle and cover layer has significant influence on the reflected and transmitted coefficients of structural surface and the energy reflection and transmitted coefficients.

Chapter 8
Prediction Model

5.12 Wenchuan earthquake induced disasters of hundreds of landslides and debris flow, and ten thousand of hazard points turned out in 39 severely affected counties solely in Sichuan Province, in which landslide hazard point accounted for 41 %, which brought huge loss to China and our people. However, the earthquake also urged domestic and overseas specialists and scholars to conduct further research on landslides prediction and promote many kinds of landslides prediction model and methods, which can be roughly summarized into the following three kinds: ① empirical statistical prediction model, which obtains statistical formula representing the relations of landslides horizontal sliding distance and its influence factors; ② deterministic prediction model, which is the physical model, based on the dynamic principle of mass, combining conservation laws of landslides motion mode, mechanism and energy and momentum together; ③ numerical analysis prediction model, which mainly uses numerical methods such as DEM and grain flow to simulate the whole landslides process so as to predict the horizontal sliding distance of landslides.

To sum up, the currently commonly used prediction method of landslides horizontal sliding distance (L) mainly considers the influence factors such as the landslides volume (V), the landslides maximum vertical sliding distance (H), and equivalent friction coefficient ($f = H/L$), but can rarely pay attention to other factors such as PGA and the geometric size of mountains while the research by Wang Xiuying et al shows that the landslides distribution has good correlation to acceleration distribution and the mountain bevel angles. At the same time, the research results of Wenchuan earthquake hazard show that bedrock and overburden slope is the most common slope structure. Therefore, this chapter will base on the field research results of Wenchuan earthquake landslides, focus on collecting and organizing the damage of bedrock and overburden slopes, and adopt the seismic ground motion attenuation model which has been modified according to measured data of Wenchuan earthquake to estimate the PGA of each landslide hazard point. Then, it will comprehensively consider totally four influence factors including mountain influence factors (mountain height H_L, bevel angle θ), ground motion influence factors (PGA and sliding mass volume V), and build the prediction model of

Y. Changwei et al., *Slope Earthquake Stability*,
DOI 10.1007/978-981-10-2380-4_8

horizontal sliding distance of bedrock and overburden slopes induced by Wenhcuan earthquake through quaternary nonlinear regression analysis, so as to provide efficient technical support to prevention and reduction of seismic landslides.

8.1 Field Investigation

This landslide investigation went into the seismic areas for several months, which mainly includes the worst-hit areas such as Beichuan, Wenchuan, and Qingchuan (see Figs. 8.1 and 8.2). Our research group has cumulatively conducted detailed geological hazard research of 4573.8 m, and in 937 hazard points of all kinds, in which 834 collapse and landslide hazard points/groups (678 collapse and 145 landslide), and 103 debris flow hazard point, and we got the valuable first hand data about Wenchuan earthquake geological hazard.

Based on the field research results, this chapter put emphasis on organizing and analysis of damage in 65 classic landslides, in which the 65 landslides were taken as statistical analysis objects, and the Donghekou landslide and Magongwoqian landslide were taken as verification objects for the correctness and advantage of the prediction model. For the 65 landslides, the maximum distance of the hazard point to fault is about 248 km and the minimum distance is about 1 km; the maximum collapse and landslide volume is about 750 million cubic meters and the minimum

Fig. 8.1 Map of landslide field in Donghekou

Fig. 8.2 Field investigation team of landslide disaster

is about 2000 m^3; the maximum bevel angle of disaster slope is about 57° and the minimum is about 19.3°; the maximum height of the mountain is about 1250 m and the minimum is about 40 m; the maximum horizontal sliding distance is about 5300 m and the minimum is about 20 m; the maximum equivalent friction coefficient is about 1.511 and the minimum is about 0.158; the maximum estimated PGA is about 1.015 g and the minimum is about 0.062 g.

8.2 Controlling Factors

Although there are multiple factors that influence the horizontal sliding distance L of bedrock and overburden slopes, the mountain influence factors (mountain height H_L, bevel angle θ), and ground motion influence factors (PGA and sliding mass volume V) are the major controlling factors, which have been widely acknowledged. Thus, the influence factors of bedrock and overburden slopes induced by earthquakes that will be determined in this chapter are mountain height H_L, bevel angle θ, PGA, and collapse and landslide volume V (see Fig. 8.3 for details).

In Fig. 8.3, H_L stands for the mountain height; V stands for volume; L is the horizontal distance between sliding mass leading edge after landslide and its trailing edge before landslide, that is, the horizontal sliding distance of landslide; θ is the bevel angle before landslide. Besides, H stands for the vertical sliding distance of landslide.

8.3 Analysis of Controlling Factors

This section focuses on the controlling factors of bedrock and overburden slopes induced by earthquakes, adopts field investigation data of bedrock and overburden slopes induced by Wenchuan earthquake, and combines domestic and overseas

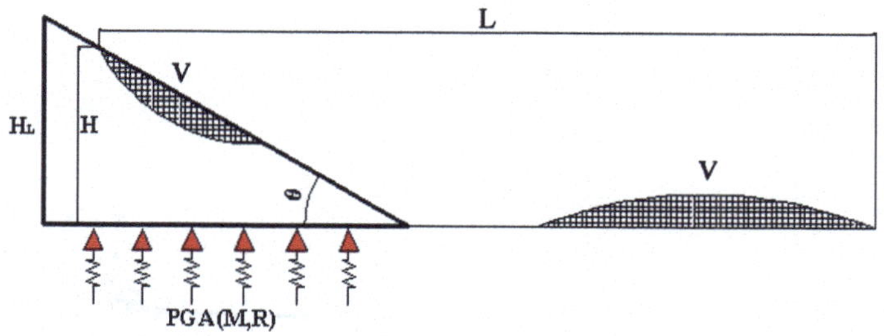

Fig. 8.3 Sketch map of landslide horizontal sliding distance influence factors

seismic disaster data of the same kind to conduct correlation research among controlling factors and the seismic geological damage range.

8.3.1 Correlation Between Landslide Volume V and Horizontal Sliding Distance L and Vertical Sliding Distance H

For the influence of landslide volume on the horizontal sliding distance of bedrock and overburden slopes, scholars from domestic and abroad have conducted wide and in-depth exploration and study, for example, Budetta et al. have carried out statistical analysis to the sliding distance and volume of 25 bedrock and overburden slopes in Southern Italy, and the results show that the horizontal sliding distance of landslides has good positive correlation to its volume. Graziella conducted research on 33 bedrock and overburden slope landslides in Central America, and the results show that the horizontal sliding distance of landslides has good positive correlation to its volume. Li Xiuzhen and Kong Jiming et al conducted research on 46 classic bedrock and overburden slope landslides induced by Wenchuan earthquake, and the statistical results show that the horizontal and vertical sliding distance of landslides grow exponentially with the landslide volume. This section conducted statistical analysis of the correlation between landslide volume *V* and the horizontal sliding distance *L* and vertical sliding distance *H* using the collected 65 classic bedrock and overburden slope landslides induced by Wenchuan earthquake, and discovered both log*V*-*L* and log*V*-*H* positive correlation of exponential function (see Figs. 8.4 and 8.5).

8.3.2 Correlation Between Bevel Angle tanθ and Equivalent Friction Coefficient *f*

Domestic and overseas specialists and scholars have conducted certain research and exploration on the influence factors of the mountain itself on the horizontal sliding distance of bedrock and overburden slope landslides. Okura et al. have verified the

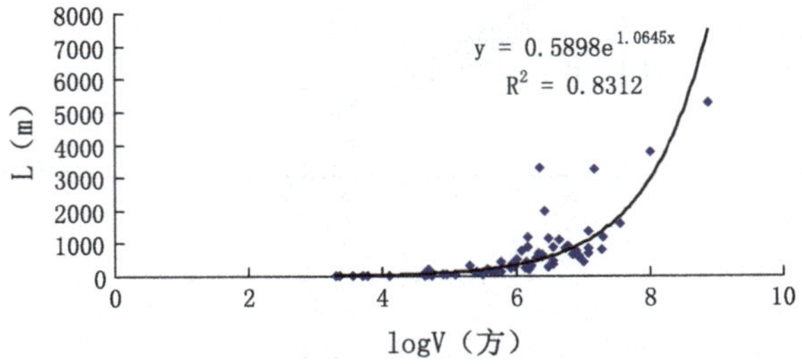

Fig. 8.4 Diagram of relation between horizontal sliding distance *L* and landslide volume

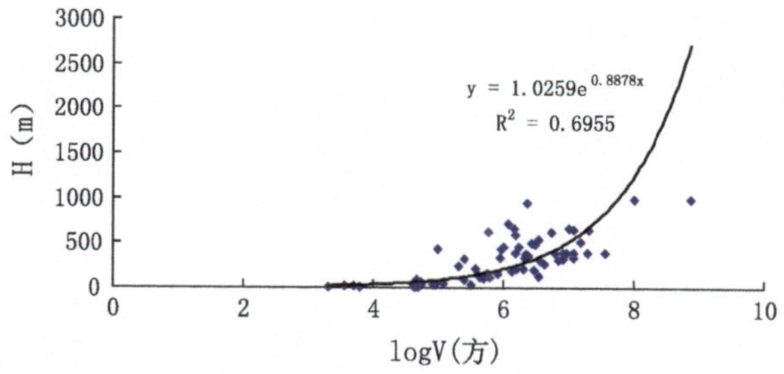

Fig. 8.5 Diagram of relation between vertical sliding distance *H* and landslide volume

positive correlation between equivalent friction coefficient f and bevel angle θ through numerical analysis simulation. Hattanji et al conducted statistical analysis of 338 landslides of bedrock and overburden slopes in four regions of Japan and found out that the equivalent friction coefficient f and bevel angle θ were in good positive correlation and the square value of correlation coefficient was in the range of 0.78–0.88. The above research demonstrates that the influence factors of the mountain itself have large influence on the horizontal sliding distance of landslides. This section carried out statistical analysis of the correlation between the tangent value of original bevel angle and the equivalent friction coefficient f with the collected data of the 65 classic bedrock and overburden slope landslides induced by Wenchuan earthquake and found the two had good linear positive correlation with the correlation coefficient = 0.74, which is close to the correlation coefficient obtained by Hattanji (see Fig. 8.6). At the same time, it can be known by the statistical analysis results of correlation between landslide position and mountain height that the probability of landslide happening at $1/3H_L$ is 70.8 %, that of landslide happening at 0.3–0.6H_L is 15.4 %, and that of landslide happening at $1/3H_L$ is 13.8 %, so most landslides happen at $1/3H_L$ (see Fig. 8.7).

8.3.3 *Seismic Ground Motion Attenuation Model Based on Measured Data of Wenchuan Earthquake*

The selection of seismic ground motion is crucial to building landslide prediction model induced by earthquake. Due to the lack of measured data of seismic ground

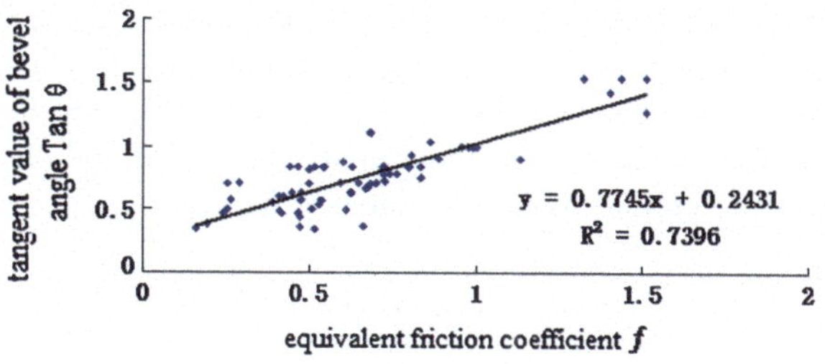

Fig. 8.6 Correlation between tangent value of bevel angle and equivalent friction coefficient f

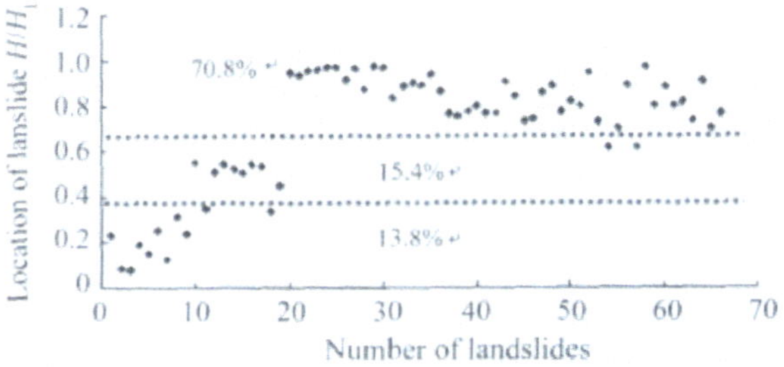

Fig. 8.7 Map of landslide position distribution

motion for long time, most of the seismic landslide prediction model in China misses the item of seismic influence factors. This section will adopt seismic ground motion attenuation model to evaluate PGA on landslide hazard points. Currently, most attenuation models are empirical relations obtained by regression of historical data. In 2006, Zhao et al. built a spectral acceleration attenuation model based on large amount of seismic records [see Fig. (8.1)]. In 2010, Lu et al. conducted comparative research of response spectrum on Wenchuan earthquake seismic prediction model, and the results showed that, compared with multiple Next Generation Attention attenuation models promoted by American scholars, the prediction of PGA and short period spectral acceleration in near-field region of Wenchuan earthquake model promoted by Zhao is more accurate. Therefore, this section will adopt the adjusted seismic ground motion attenuation model promoted by Zhao based on measured data of Wenchuan earthquake to obtain PGA of unknown landslide hazard points. The concrete method is as follows: First, use this model to work out the PGA prediction value of 131 test stations within Sichuan Province, compare the results with the measured seismic records of seismic array station, and work out the logarithmic residual distribution of the two with the fault distance (see delta point in Fig. 8.8); second, adopt linear regression method to calculate the linear trend line of this distribution (see the dotted line in Fig. 8.8), and see Formula (8.2) for its function.

$$\begin{cases} \ln \mathrm{PGA} = \ln(f(M,R)) = aM_{wi} + bx_{i,j} - \ln(r_{i,j}) + e(h-h_c)\delta_h + F_R + S_L + S_S + \\ S_{SL}\ln(x_{i,j}) + C_k + \xi_{i,j} + \eta_i \\ r_{i,j} = x_{i,j} + c\exp(dM_{wi}) \end{cases} \tag{8.1}$$

$$y = 0.001242x - 0.059721 \tag{8.2}$$

For the relevant parameters in Formula (8.1), see reference. In Formula (8.2), x stands for the fault distance (unit: km), and y stands for the log residuals of PGA measured value and its prediction value. Bringing the above formula into seismic ground motion attenuation model promoted by Zhao, we will get the adjusted seismic ground motion attenuation model.

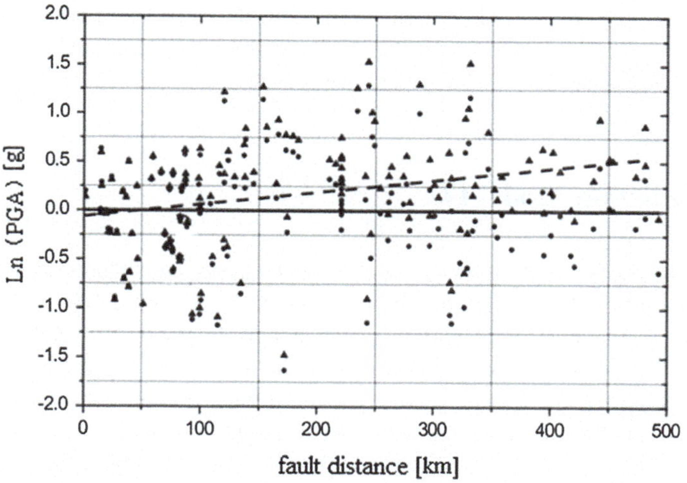

Fig. 8.8 Logarithmic residual distribution of Zhao model with fault distance

8.3.4 *Correlation Between PGA and Horizontal Sliding Distance*

Based on the latitude/longitude coordinates and fault position of each hazard point of bedrock and overburden slopes, we worked out the correspondent minimum fault distance and adopted the adjusted seismic ground motion attenuation model to work out PGA of each hazard point of bedrock and overburden slopes, and further studied on the influence of PGA on the horizontal sliding distance of bedrock and overburden slopes (see Fig. 8.9 for detailed results).

Through comprehensive analysis of Fig. 8.9, it can be known that direct proportion linear relation (R^2 = 0.4037) exists between PGA and horizontal sliding distance of bedrock and overburden slopes log*L*, i.e., horizontal sliding distance extends with the PGA increase. Therefore, PGA grows exponentially with the

Fig. 8.9 Diagram of correlation of PGA and horizontal sliding distance

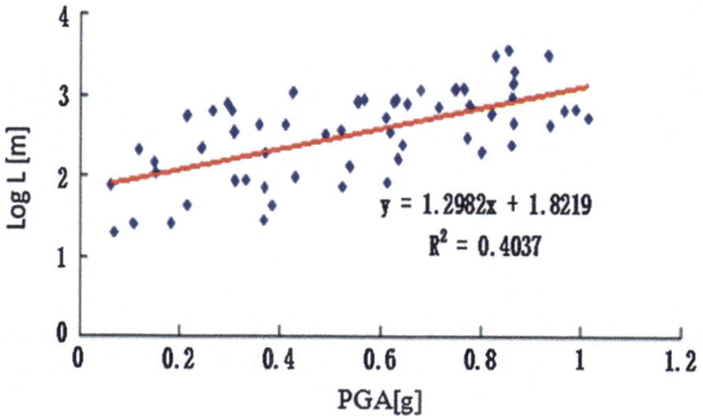

horizontal sliding distance. It is noteworthy that the correlation coefficient R^2 of PGA and log$L$ is low, which may be because the test points are few and they distribute widely, so there is large data discreteness.

8.4 Prediction Model O

Based on the research results of Sect. 8.3, it can be known that the horizontal sliding distance L has good exponential function correlation with PGA and landslide volume V; as the horizontal sliding distance L is in good exponential function correlation with vertical sliding distance H, and H is in good linear correlation with mountain height H_L, so L is in good exponential function correlation with $H_L + A$, in which A is the constant; similarly, we can get that logL is in good linear correlation with (tanθ − B), in which b is the constant. Therefore, this section will conduct quaternary linear regression of the above four factors so as to work out the sliding distance prediction model of bedrock and overburden slopes induced by Wenchuan earthquake.

$$\log L = \alpha \log(H_L + A) + \beta \log V + \delta \log(\tan\theta - B) + \mu f(M,R) + \eta \qquad (8.3)$$

In the Formula, L stands for horizontal distance, i.e., damage (m); V stands for landslide volume (m^3); H_L stands for mountain height (m); θ stands for bevel angle (°); $f(M, R)$ represents PGA (g); η stands for constant, in which, H_L and θ represent the geometric factors of the slope; f (M, R) represents the seismic ground motion geometric factor and it is worked out by seismic ground motion attenuation model promoted by Zhao.

Carrying out regression analysis, we got $\alpha = 0.2018$, $\beta = 0.3699$, $\delta = -0.1132$, $\mu = 0.3598$, $\eta = -0.40922$, $A = -40$, $B = 0.325$, and the standard deviation of log$L$ is 0.026. For the residual plot of the regression equation, see Figs. 8.10, 8.11, 8.12, and 8.13.

It can be known from Figs. 8.10, 8.11, 8.12, and 8.13 that the linear trend line of the log(H_L + A), PGA, log(tanθ – B), and log$V$ residual basically keep horizontal, approaching the O axle. Through comprehensive analysis of prediction model of

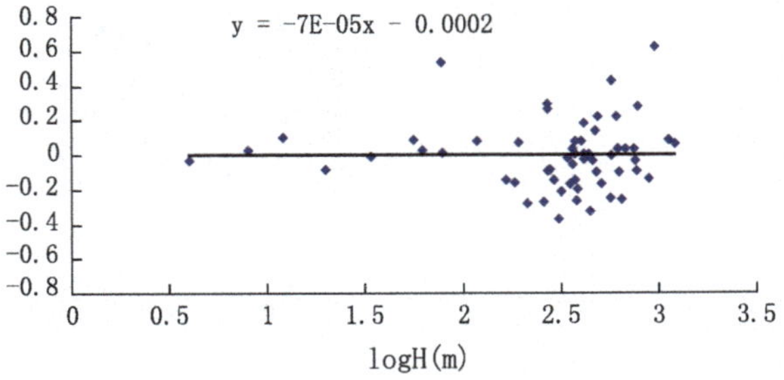

Fig. 8.10 Residual plot of Log*H*

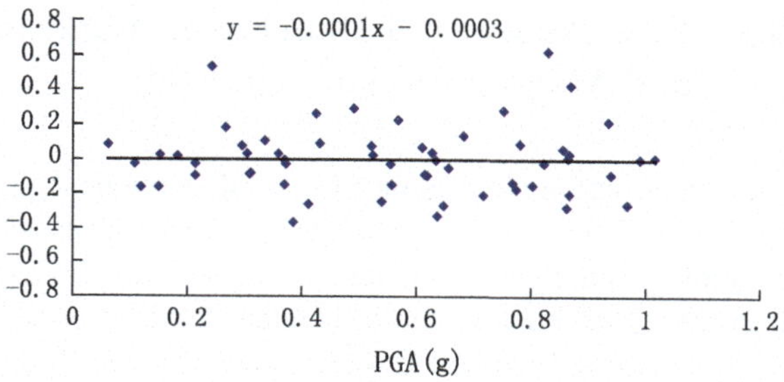

Fig. 8.11 Residual plot of PGA

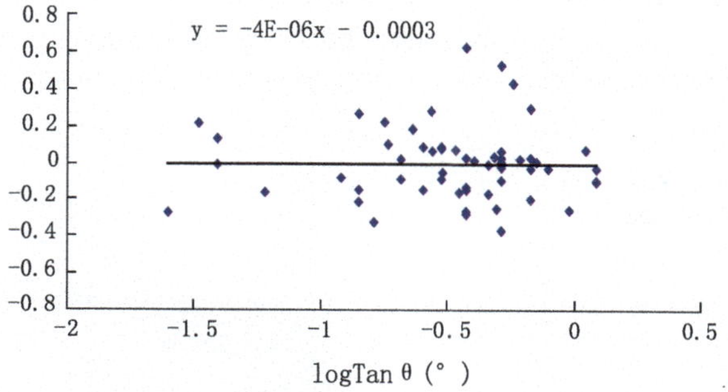

Fig. 8.12 Residual plot of Log(tanθ)

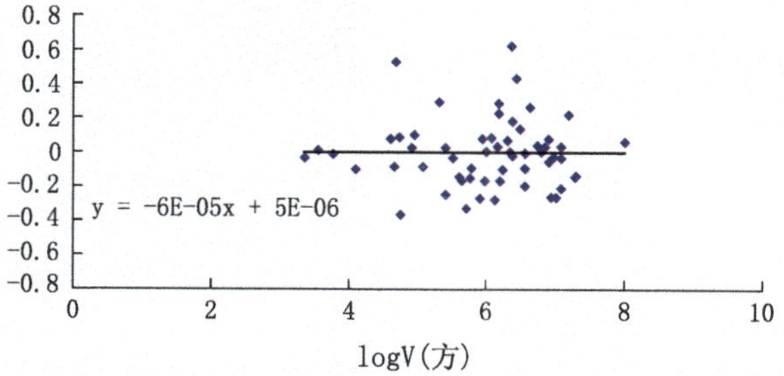

Fig. 8.13 Residual plot of Log$V$

horizontal sliding distance of bedrock and overburden slopes induced by Wenhcuan earthquake, it can be known that, when H_L, θ, PGA, and V are known, we can predict the probable horizontal sliding distance of bedrock and overburden slopes, so as to decide the landslide damage range and provide scientific reference to decision on scheme of relocation and preventing collision, engineering management, and relocation range.

8.5 Case Analysis of Donghekou Landslide and Magongwoqian Landslide

Wenchuan earthquake induced lots of geological disasters such as collapse and landslides that formed large amount of barrier lakes and caused serious damage. For example, Donghekou landslide situated in Donghekou Village, Hongguang Country, Qingchuan County, Sichuan Province, which respectively formed barrier lakes of one million square meters and three million square meters, and seriously jeopardized people's life and property safety (see Fig. 8.14). At the same time, Magongwoqian landslide in Qingchuan County was high-speed and long-runout landslide induced by Wenchuan earthquake, whose volume was 12 million square meters and it caused 38 casualties (see Fig. 8.15).

This section takes the above two landslides as example, adopts the adjusted Wenchuan seismic ground motion attenuation model to estimate the PGA of these

Fig. 8.14 Donghekou landslide

Fig. 8.15 Magongwoqian landslide

and gets 0.934 and 0.956 g, and combines with the field research results to verify the correctness of the sliding distance prediction model of bedrock and overburden slopes induced by Wenhcuan earthquake. The field research results of our research group are basically the same with others (see Table 8.1). At the same time, we compare the calculation results in the above model with the empirical method of traditionally used "estimate the landslide damage range with twice of the altitude difference of slope leading edge and trailing edge" and the calculation results of Sen Xiekuan method to further study on the advantage of the above prediction model. See Table 8.2 for detailed results.

Table 8.1 Research results of relevant parameters of Donghekou landslide and Magongwoqian landslide

Parameters	Donghekou landslide			Magongwoqian landslide	
	Our research group	Sun Liping et al.	Fu Rong et al.	Our research group	Zhang Wei et al.
Mountain height (m)	1610	1500	1330	1950	1870
Average bevel angle (°)	41	45	39	38	35
Landslide volume (106 m^3)	15	/	15	12.1	12
Horizontal sliding distance (m)	2200	2400	2450	1880	1960

Table 8.2 Comparison of horizontal sliding distance of Donghekou landslide and Magongwoqian landslide with different methods

	The measured value	The prediction model	The empirical model	Sen Xiekuan method ($\Delta H_1/L = 0.73\tan\alpha - 0.07$)
Donghekou landslide				
Horizontal sliding distance (m)	2200	2023	960	921.8
Absolute value of error (m) (%)		8.05	56.36	58.1
Magongwoqian landslide				
Horizontal sliding distance (m)	1880	1960	1400	1469
Absolute value of error (m)		4.26	25.53	21.86

Note ΔH1 stands for the altitude difference of leading edge and trailing edge of the slope after landslides; α is the bevel angle before landslides

Through comprehensive analysis of Table 8.2, it can be known that the calculation results of Sen Xiekuan method and empirical method differ considerably. It may be because the geological conditions and the external environment needed by the formation of the methods are quite different from Wenchuan earthquake; so the prediction accuracy is low; the prediction errors of the prediction model of horizontal sliding distance of bedrock and overburden slopes induced by Wenhcuan earthquake are 8.05 % (Donghekou landslide) and 4.26 % (Magongwoqian landslide) and its prediction accuracy is high. Therefore, the prediction model proposed in this dissertation has high prediction accuracy and can rationally predict the horizontal sliding distance of landslides.

8.6 Parameters

Based on the seismic induced prediction model of horizontal sliding distance of bedrock and overburden slopes, this section will select relevant parameters to study the influence of V, tanθ, H_L, and PGA on the horizontal sliding distance of bedrock and overburden slopes, and the details are as follows: ① When landslide volumes are ten million m^2, five million m^2, two million m^2, one million m^2, 500 thousand m^2, 200 thousand m^2, and 10 thousand m^2, the bevel angle is 30°, the mountain height is 1000 m, and PGA is 0.2 g, work out the influence of landslide volume on horizontal sliding distance L; ②when the landslide volume is one million m^2, the H_L is 1000 m, PGA is 0.2 g, and the bevel angles are 20°, 30°, 40°, and 45°, work out the influence of bevel angle θ on horizontal sliding distance L; ③ when V is one million m^2, PGA is 0.2 g, the bevel angle is 30°, and the H_L are 1000, 800, 600, 400, 200, and 100 m, work out the influence of H_L on horizontal sliding distance L; ④ when V is one million m^2, H_L is 1000 m, the bevel angle is 30°, and PGA are 0.2, 0.4, 0.6, 0.7, 0.8, and 1.0 g, work out the influence of PGA on horizontal sliding distance L. For the detailed calculation results of above situations, see Figs. 8.16, 8.17, 8.18, and 8.19.

Fig. 8.16 Correlation curve of L–tanθ

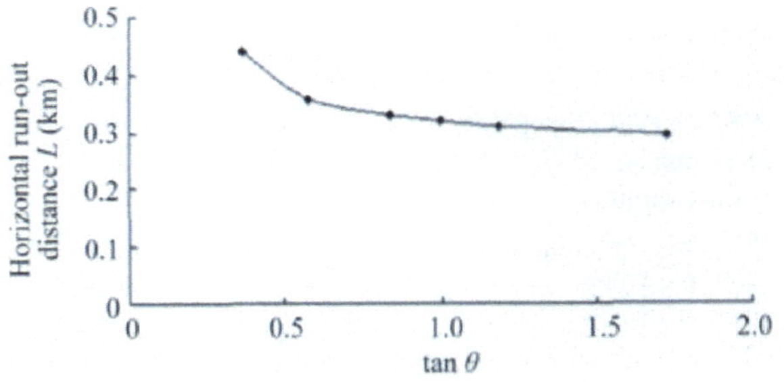

Fig. 8.17 Correlation curve of *L*–PGA

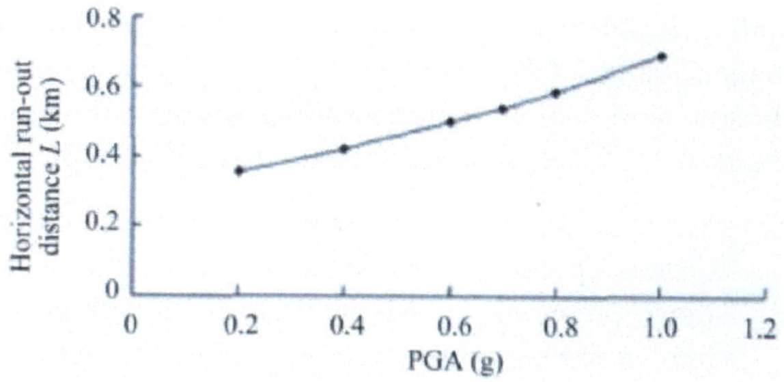

Fig. 8.18 Correlation curve of *L*–*H*

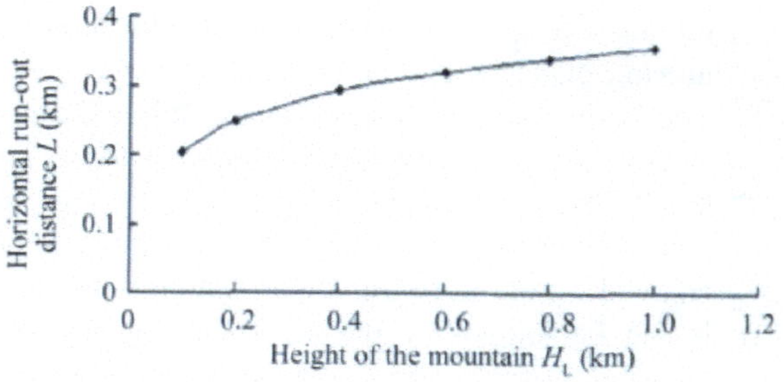

Fig. 8.19 Correlation curve of *L*–*V*

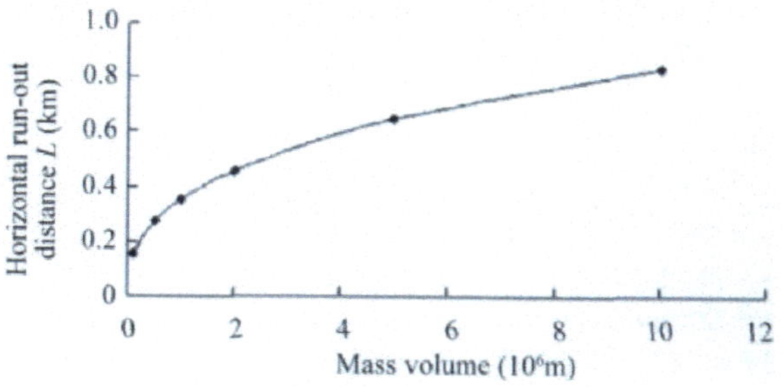

Through comprehensive analysis of Figs. 8.16, 8.17, 8.18, and 8.19, it can be known that L is in positive correlation with PGA, H_L, and V, and in negative correlation with $\tan\theta$, which is basically identical with the results of statistical analysis of the field research results of Wenchuan earthquake landslides.

8.7 Brief Summary

Mountain height H_L, bevel angle θ, PGA, and landslide volume V are the four important factors influencing the horizontal sliding distance of bedrock and overburden slopes. On the basis of detailed research on landslide damage of Wenchuan earthquake, we conducted statistical analysis results of correlation of the horizontal sliding distance L of bedrock and overburden slopes with mountain height H_L, tangent value of bevel angle $\tan\theta$, PGA, and landslide volume V, built the prediction model of horizontal sliding distance of bedrock and overburden slopes induced by Wenhcuan earthquake, and came to the following conclusions:

(1) Select Donghekou large scale landslide and other two commonly used sliding distance prediction model to verify the rationality and advantage of damage range prediction model of Wenchuan earthquake-induced bedrock and overburden slopes, which fully proved its feasibility. At the same time, the model can not only predict horizontal sliding distance of bedrock and overburden slopes, but also can decide the landslide damage range and provide scientific reference to decision on scheme of relocation and preventing collision, engineering management, and relocation range.
(2) The prediction model of horizontal sliding distance of bedrock and overburden slopes can take good consideration of PGA, landslide volume V, mountain height H_L, and tangent value of bevel angle $\tan\theta$, four influence factors in total. The parameter research shows that L is in positive correlation with PGA, H_L, and V, and in negative correlation with $\tan\theta$.
(3) It can be known by the statistical analysis results of correlation between landslide position and mountain height that the probability of landslide happening at $1/3H_L$ is 70.8 %, that of landslide happening at 0.3–$0.6H_L$ is 15.4 %, and that of landslide happening at $1/3H_L$ is 13.8 %, Therefore, most landslides happen at $1/3H_L$.

Conclusions and Prospects

The slope stability under seismic effects is a both ancient and young subject. Because of its high complexity, the development till now is still immature. The special geographical and geological conditions and seismic configuration make it especially pertinent in China. Currently, the infrastructure construction such as highway, railway, and water conservancy is in rapid developing period, which draws higher demand on seismic design of slope engineering. Therefore, the slope stability under seismic effects is a crucial scientific problem urgently needs to be solved.

Part One: The Major Work and Conclusion of the Dissertation

Surrounding the national demands on infrastructure construction and reconstruction of Wenchuan seismic areas, and based on the investigated data of Wenchuan seismic slope damage, the research results of seismic array monitoring, the large-scale shaking table test and the numerical analysis, and technical measures like theoretical analysis, we explored the dynamic features of rock slope. On this basis, we take the most seriously hit bedrock and overburden slope in the earthquake as study object. This dissertation systematically studied on the scientific issues such as the rock slope dynamic response rules under seismic effects, the formation mechanism of landslide, and stability evaluation method as well as landslide hazard evaluation, which covered problems of slope stability under seismic effects on micro- and macro-aspects, and achieved some results and conclusions in the following aspects.

On Basic Dynamic Response of Rock Slope

(1) Based on the seismic array monitoring results of Zigong topography in "5.12 Wenchuan earthquake," we conducted research on the seismic dynamic characteristics of slope topography from the aspects of amplitude response,

Y. Changwei et al., *Slope Earthquake Stability*,
DOI 10.1007/978-981-10-2380-4

Fourier spectrum, and response spectrum and achieved the following results: The component PGA of each direction on slope topography increases in nonlinear trend along elevation, the acceleration elevation amplification effects on EW direction were stronger than those on NS direction and those on UP direction were the weakest; for the horizontal acceleration, with the increase of elevation, the frequency components of Fourier spectrum around the slope natural frequency had amplification effects and showed the changing process from single peak to double peak, while the Fourier spectrum of vertical acceleration basically stayed unchanged; with the increase of elevation, the dynamic amplification coefficient of acceleration response spectrum increased, so did amplification effects, and the dynamic amplification coefficient of horizontal acceleration response spectrum was more than that of vertical acceleration response spectrum (Chap. 2).

(2) This chapter based on the large-scale shaking model test studied on the influence of bevel angle, seismic intensity, and seismic wave input type on rock slope dynamic characteristics, and its results showed: For mountains containing four slope surfaces, if we analyze with traditional method, i.e., using X, Y, and Z three direction coordinate system to analyze PGA elevation amplification effects, the results may be unilateral, so it suggested to use the three-dimensional local coordinate system. i.e., the free face direction, slope strike direction, and vertical direction of slopes; with the increase of bevel angles, the PGA elevation amplification effects in vertical direction will rise gradually, while at the same time, there exists a sudden increase turning point at bevel angles of 45° and a leveling off turning point at 50°. However, the PGA elevation amplification effects in slope strikes stay unchanged with the increase of bevel angles, and its steps are comparatively gentle; with the increase of seismic ground motion PGA, the peak acceleration elevation amplification effect decreases gradually, slope strike direction and vertical direction reduce gradually and show characteristics of magnitude saturation; the changes of Fourier spectrum frequency components of acceleration in free face direction (L), slope strike (M), and vertical direction (N) along elevation have certain regularity, i.e., with the increase of elevation, the slope soil has significant amplification effects for frequency components around its natural frequency range f and has filter effects for other frequency ranges; the changes of Fourier spectrum frequency components of acceleration in free face direction (L), slope strike (M), and vertical direction (N) along elevation are basically identical, and the correspondent response spectrum amplitude of predominant period has certain amplification effects along elevation, while for other period T, especially long period parts(low-frequency parts), there exists certain reduction effects. In addition, it has obvious peak phenomena for free face direction, while for slope strike direction and vertical direction, the peak phenomena are less obvious (Chap. 3).

(3) This chapter adopted both shaking table test and numerical analysis, and studied the influence of factors such as slope type, slope top local topography on the rock slope dynamic characteristics under seismic effects from the

perspective of time domain and joint time–frequency domain, and the results show: The acceleration Fourier spectrum along elevation of single- and double-sided high and steep slope surfaces all shows the changing rules of low-frequency components amplification, and the changing rules are influenced by slope topography and bevel angles. No matter for horizontal acceleration or vertical acceleration, the acceleration response spectrum along elevation of single-sided slope surfaces all shows single peak value, while that of double-sided slope surfaces followed the process of converting from single peak value to double peak value; no matter for single-sided high and steep slope or for double-sided high and steep slope, the PGA of different positions along elevation all amplifies to different extent, and it shows as follows: The amplification effects of double-sided high and steep slope on acceleration > the amplification effects of single-sided high and steep slope on acceleration; PGA amplification effects on slope surface > PGA amplification effects inside slope mass; the horizontal acceleration coefficient > the vertical acceleration coefficient, and acceleration coefficient on steep slope topography > acceleration coefficient on gentle slope topography; the local topography of slope top has significant influence on the intensity of seismic response, and the detailed rules are as follows: three peak slope top > double peak slope top > single peak slope top > flattop; and the high-frequency components of all of them gradually strengthen and the second step frequency becomes gradually obvious. At the same time, the characteristic periods of acceleration response spectrum of flattop, single, double, and three peak slopes are gradually transferred from long period to short period, and the double peak value of response spectrum gradually becomes obvious (Chap. 4).

On Theoretical Analysis of Rock Slope Acceleration Elevation Amplification Effects

(4) The double-sided rock slope acceleration elevation amplification effects time–frequency method is derived, based on elastic wave theory and horizontal slice method, through building force equilibrium differential equation of micro-unit, using Hilbert–Huang Transform and boundary conditions, and applying modal analysis method and normal mode theory. The rationality of this method is verified through shaking model test and numerical analysis. This method could not only take good consideration of the influence of three essential elements (PGA, frequency, and duration) on acceleration elevation amplification, but also explain the "whiplash effects" of slopes in the perspective of theoretical calculation. With the increase of seismic intensity, the shear displacement at slope top gradually increases, and the acceleration amplification coefficient gradually decreases; the shear displacement of slope mass is in nonlinear distribution along elevation and suddenly increases in multiple times; with the

increase of input wave frequency, the maximum shear displacement and acceleration amplification coefficient at slope top are distributed in saddle shape and reach the maximum value when input frequency value equals the maximum value of the natural frequency of slope mass; with the increase of bevel angles, the maximum shear displacement and acceleration amplification coefficient all gradually increase, and distribute in "step-form," i.e., in the ranges of 30°–45° and 50°–60°, the amplification coefficient all increases slightly, while in the range of 45°–50°, it increases suddenly. Through comparative analysis of high and steep rock slope time–frequency analysis method, with the calculation results of acceleration elevation amplification effects analysis method regulated in Code for Seismic Design of Buildings GB50011-2010, Specifications for Seismic Design of Hydraulic Structures DL/5073-2000, and Design Specification for Slope of Hydropower and Water Conservancy Project SL/386-2007, it is suggested that the calculation of rock slope acceleration elevation amplification coefficient in Seismic Code, Hydraulic Seismic Specifications, and Slope Design Specifications should take consideration of the influence of seismic intensity, so as to improve the rationality of structural seismic design (Chap. 5).

On Slope Deformation Characteristics and Formation Mechanism of Bedrock and Overburden Slopes

(5) This chapter focused on the two major landslide hazard spots on the left side of G213, built the numerical analysis model based on the shaking table test model, and used the new numerical calculation software—GDEM to simulate the whole process of deposit body landslides of single- and double-sided bedrock and overburden slopes from deformation to failure slip under intense earthquake, conducting comprehensive and in-depth research of slope deformation features and formation mechanism of bedrock and overburden slopes from the perspective of time domain and joint time–frequency domain. (1) The results show: No matter for single- or double-sided bedrock and overburden slopes, the landslides processes are basically the same, i.e., under effects of gravity and seismic power, tensile stress concentration firstly occurs at sliding mass top, causing deformation of sliding mass along sliding mass structural surface trailing edge, which causes tensile and shear failure points at this position. Then, with continuous seismic power, the shear failure points on sliding mass structural surface gradually develop to lock-fixed section of sliding mass leading edge with increasing amount of tensile failure points on sliding mass surface, which finally causes progressive damage of lock-fixed section, crack surface connecting to sliding zone and sliding mass slipping from shear crack into landslide; the inconsistency of motion between sliding bed and sliding mass, difference of seismic energy distribution and dissipation,

and decrease of sliding mass structural surface strength with increase of seismic intensity are three major controlling factors inducing landslides. When seismic ground motion acceleration is small, the instant frequency of horizontal acceleration inside the sliding bed and sliding mass stabilizes in a certain range with the former slightly higher than the latter. At the same time, the energy transmission coefficient on the sliding mass structural surface and the controlling frequency range of seismic energy inside the sliding mass all stabilizes in a certain range; with the increase of seismic intensity, the instant frequency of acceleration inside the sliding mass gradually decreases, and the energy transmission coefficient of sliding mass structural surface gradually decreases, with both finally approaching stability. Simultaneously, the controlling frequency range of seismic motion energy inside the sliding mass transferrs from high-frequency range to low-frequency range (Chap. 6).

On Seismic Stability Time–Frequency Analysis Method of Bedrock and Overburden Slope

(6) Based on elastic wave theory and general geological analysis model, this chapter derived seismic stability time–frequency analysis method of bedrock and overburden slope, and verified its rationality through shaking table test and numerical analysis results. This method could not only consider the influence of three essential elements (PGA, frequency, and duration) of SV wave on slope stability, but can also predict the safety of rock slopes, the happening time, and scale of landslides. At the same time, it can provide reference to supporting structure seismic time–frequency design in high intensive seismic regions; at the same time, the elasticity modulus of incident angle and cover layer has significant influence on the reflected and transmitted coefficients of structural surface and the energy reflected and transmitted coefficients (Chap. 7).

On Damage of Bedrock and Overburden Slope Landslide Induced by Earthquake

(7) On the basis of detailed research on landslide damage of Wenchuan earthquake, we focused on collecting and organizing the damage of bedrock and overburden slopes induced by earthquakes, and conducted statistical analysis results of correlation of the horizontal sliding distance L of 65 classic landslides with mountain height HL, tangent value of bevel angle $\tan\theta$, PGA, and landslide volume V, built the prediction model of horizontal sliding distance of bedrock and overburden slopes induced by Wenhcuan earthquake, and also

selected Donghekou large-scale landslide and other two commonly used sliding distance prediction model to verify the rationality and advantage of damage range prediction model of Wenchuan earthquake-induced bedrock and overburden slopes. The research results show: L is in positive correlation with PGA, HL, and V, and in negative correlation with tanθ; the probability of landslide happening at 1/3HL is 70.8 %, that of landslide happening at 0.3–0.6HL is 15.4 %, and that of landslide happening at 1/3HL is13.8 %, Therefore, most landslides happen at 1/3HL (Chap. 8).

(1)–(6) belong to research on micro-aspects, whose conclusion could be used to seismic design of concrete slope engineering; (7) belongs to macro-aspects, whose conclusion could be applied to decision on scheme of relocation and preventing collision, engineering management, and relocation range.

Part Two: Creativity of This Dissertation

The creativity of this dissertation is as follows:

(1) Tit proposed the double-sided rock slope acceleration elevation amplification effects time–frequency method, which can not only take good consideration of the influence of three essential elements (PGA, frequency, and duration) on acceleration elevation amplification, but also can provide theoretical supports to seismic design of rock and soil engineering in mountain regions in China (Chap. 5).

For slope deformation features and formation mechanism of bedrock and overburden slopes, this chapter conducted in-depth research from the perspective of joint time–frequency domain, promoted new analysis ideas to seismic mechanism of rock and soil engineering, which broke through the traditional analysis with sole reference to time-domain features or frequency-domain features (Chap. 6).

(2) This chapter promoted seismic stability time–frequency analysis method of bedrock and overburden slope, which could consider the influence of three essential elements (PGA, frequency, and duration) of seismic wave on slope stability (Chap. 7),

(3) This chapter promoted damage range prediction model of Wenchuan earthquake-induced bedrock and overburden slopes, which could consider four key factors including mountain height HL, tangent value of bevel angle tanθ, PGA, and landslide volume V, and it has high accuracy, and could provide scientific reference to decision on scheme of relocation and preventing collision, engineering management, and relocation range (Chap. 8).

Part Three: Problems and Prospects

This dissertation conducted some beneficial exploration on the aspects of slope dynamic characteristics, dynamic response, seismic design method, landslide mechanism, stability evaluation method and landslide damage range evaluation, and obtained some creative results. But as stated in the beginning of this dissertation, the slope dynamic stability problem is a complex subject concerning multi-crossed disciplines. Due to limits of knowledge of the author and research conditions, there is still imperfection in this dissertation, and there are problems worth further study and discussion.

(1) Conducting research on landslide damage mechanism, failure mode, motion features, and damage range;
(2) Conducting further study on constitutive relation of rock and soil mass, and building correspondent constitutive relation model of rock and soil mass based on different materials and loading effects, which has important value to both theoretical analysis and engineering application, and is a key direction for research of slope dynamic stability analysis;
(3) Conducting research on deformation mode and long-term evolutionary mechanism of rock slopes;
(4) Conducting research on slope stability judging system based on nonlinear characteristics of slope structural surface;
(5) Conducting research on theory of acceleration elevation amplification effects with full consideration of nonlinear characteristics of rock and soil mass.

Bibliography

Ai-Homoud AS, Tahtamoni WW. Reliability analysis of three-dimensional dynamic slope stability and earthquake-induced permanent displacement. Soil Dyn Earthq Eng. 2000;19(2):91–114.

Ausilio E, Conte E, Dente G. Seismic stability analysis of reinforced slopes. Soil Dyn Earthq Eng. 2000;19(3):159–72.

Bai GL, Xue F, Xu YZ. Seismic damage analysis and reduction measures of buildings in village and town in the Yushu Earthquake. J Xi'an Univ Archit Technol (Natural Science Edition). 2011;43(6):309–15.

Biondi G, Cascone E, Maugeri M. Flow and deformation of sandy slopes. Soil Dyn Earthq Eng. 2002;22(10):1103–14.

Bo JS, Xu GD, Jing LP. Seismic response and dynamic stability analysis of soil slopes. Earthq Eng Eng Vib. 2001;21(2):116–20.

Bray JD, Repetto PC. Seismic design considerations for lined solid waste landfills. Geotext Remembr. 1994;13(8):497–518.

Budetta P, De Riso R. The mobility of some debris flows in pyroclastic deposits of the northwestern Campanian region (southern Italy). Bull Eng Geol Environ. 2004;63:293–302.

Cai SH, Wang LM, Yuan ZX. A Preliminary study on the seismic landslide distance in the shanxi-gansu-ningxia-shanxi Loess region. Northwest Seismol J. 1998;20(4):75–82.

Celebi M. Topographic and geological amplification determined from strong-motion and aftershock records of 3 March 1985 Chile earthquake. Bull Seismol Soc Am. 1987;77:1147–57.

Chen XL, Ran HL, Wang MM. Hazards zonation for potential earthquake-induced landslide area. Chin J Geophys. 1999;35(6):24–5.

Chen ZY, Mi HL, Wang XG. A three-dimensional limit equilibrium method for slope stability analysis. Chin J Rock Mech Eng. 2001;23(5):525–9.

Chen YM, Ke H, Ling DS. Dynamic properties and seismic response of municipal solid waste. China Civ Eng J. 2002;35(3):66–72.

Chen LL, Chen MZ, Qian SG. Stability analysis of high-steep rocky slope under earthquake loads. J Yangtze River Sci Res Inst. 2004;21(1):33–5.

Clough RW, Chopra AK. Earthquake stress analysis in earth dams. J Eng Mech, ASCE. 1966;92 (EM2).

Crespellani T, Madiai C, Vannucchi G. Earthquake destructiveness potential factor and slope stability. Geo-technique. 1998;48(3):411–9.

Cui P, Wei FQ, Chen XQ. Geo-hazards in Wenchuan Earthquake Area and Countemeasures for Disaster Reduction. Bull Chin Acad Sci. 2008;23(4):317–23.

Devoli G, De Blasio FV, Elverh A. Statistical analysis of landslide events in Central America and their run—out distance. Geotech Geol Eng. 2009;27:23–42.

Ding YH, Wang YQ, Sun JZ. Correlation between landslides and seislides and seismic parameters and its application in predicting slope earthquake disaster. Chin J Geophys. 1999;42 (Suppl):101–7.

Y. Changwei et al., *Slope Earthquake Stability*,
DOI 10.1007/978-981-10-2380-4

Ding YH, Wang YQ, Sun JZ, Tang Y. Research on the method for prediction of earthquake-induced landslides and its application to engineering projects. J Eng Geol. 2000;8(4):475–80.

Du XL, Han JY, Li LY. Selcetion of shaking table test similarity Relations for Long-distance Buried Pipeline. J Disaster Prev Mitig Eng. 2013;33(3):246–52.

Duncan JM. State of the art: limit equilibrium and finite-element analysis of slopes. J Geotech Eng. 1996;22(7):577–96.

Fan XY, Qiao JP. Influence of landslide and ground factors on large-scale landslide movement. Chin J Rock Mech Eng. 2010;29(11):2337–47.

Feng WK, Xu Q, Huang RQ. Preliminary study on mechanical mechanism of slope earthquake-induced deformation. Chin J Rock Mech Eng. 2009;28(Suppl):3124–30.

Hattanji T, Moriwaki H. Morph metric analysis of relic landslides using detailed landslide distribution maps: implications for forecasting travel distance of future landslides. Geomorphol. 2009;103:447–54.

Hartzell SH, Carver DL, King KW. Initial investigation of site and topographic effects at Robin wood ridge, California. Bull Seis Soc Am. 1994;84:1336–49.

Hong YS, Chen RH, Wu CS. Shaking analysis of steep nailed slopes. Can Geotech J. 2005;42 (5):1264–79.

Hu YX, Zhang YS, Liang JW, HHT-based identification of site liquefaction. China Civ Eng J. 2006;39(2):66–77.

Huang NE, Shen Z, Long SR. The empirical mode decomposition and Hilbert spectrum for nonlinear and non-stationary time series analysis. Proc R Soc Lond. 1998; 454:903–95.

Huang RQ, Li WL. Research on development and distribution rules of geohazards induced by Wenchuan earthquake on 12th May, 2008. Chin J Rock Mech Eng. 2008;27(12):2585–92.

Hungr O. A model for the run-out analysis of rapid flow slides, debris flows, and avalanches. Can Geotech J. 1995;32:610–23.

Jiang XY, Qiao JP. Contact mechanics model for risk predication of typical landslides. Eng Mech. 2006;23(8):106–9.

Keefer DK. Landslides caused by earthquake. Bull Geol Soc Am. 1989;95:406–21.

Keefer DK. Statistical analysis of an earthquake-induce landslide distribution-the 1989 Loma Prieta, California event. Eng Geol. 2000;58(3–4):231–49.

Kurita RR, Rodriquez-Ovejero LA. Model with non-reflecting boundaries for use in explicit soil-structure interaction analyses. Earthq Eng Struct Dyn. 1980;8:361–74.

Lam L, Fredlund DG. A general limit equilibrium model for three-dimensional slope stability analyses. Can Geotech J. 1993;30(6):905–19.

Lee KL. Seismic permanent deformations in earth dams. Los Angeles: School of Engineering and Applied Science, University of California. 1974.

Leshchinsky D, Ching SK. Pseudo-static stability of slopes: Design. J Geotech Eng ASCE. 1994;120(9):1514–32.

Li SH, Liu TP, Liu XY. Analysis method for landslide stability. Chin J Rock Mech Eng. 2009;28:3309–24.

Lin JS, Whitman R. Earthquake induced displacements of sliding blocks. J Geotech Eng. 1986;112 (1):44–59.

Lin M.-L, Wang K.-L. Seismic slope behavior in a large-scale shaking table model test. Eng Geol. 2006;86(2–3):118–33.

Ling HI, Cheng AD. Rock sliding induced by seismic force. Int J Rock Mech Min Sci. 1997;34 (6):1021–9.

Liu LP, Lei ZY, Zhou FC. The evaluation of seismic slope stability analysis methods. J Chnongqing Jiaotong Univ. 2001;20(3):83–88.

Liu HS, Bo JS, Liu DS. Review on study of seismic stability analysis of rock-soil slopes. Earthq Eng Eng Vib. 2005;25(1):164–71.

Liu HS, Bo JS, Liu DD. Development on study of seismic stability evaluation methods of rock-soil slopes. J Inst Disaster Prev Sci Technol. 2007;9(3):20–27.

Lu YX, Shi YC, Chen YM. Slippage estimation of the loess landslide triggered by earthquake. Northwest Seismol J. 2006;28(3):248–51.

Lu M, Li XJ, An XW. A comparison of recorded response spectra from the 2008 Wenchuan, China, earthquake with modern ground-motion prediction models. Bull Seismol Soc Am. 2010;100(5B):2357–80.

Lveda SA, Sepu A, Murphy W, Randall C, Jibson W, et al. Seismically induced rock slope failures resulting from topographic amplification of strong ground motions: The case of Pacoima Canyon, California. Eng Geol. 2005;80(3–4):336–48.

Men YM, Peng JB, Li XC. Research on vibration testing of models for dynamic stability of rock slope with layered structures. World Earthq Eng. 2004;20(4):131–36.

Mu CX, Yan WM, Zhou Q. The application progress of horizontal isolation devices for museum free-standing cultural relics. J Water Resour Archit Eng. 2014;5(2):1–6.

Newmark NM. Effects of earthquakes on dams and embankments. Geo-technique. 1965;15 (2):139–60.

Okura Y, Kitahara H, Sammori T. The effects of rock-fall volume on run-out distance. Eng Geol. 2000;58:109–24.

Qi SL, Qi SW, Wu FQ. On permanent displacement of earthquake induced slide based on residual pushing force method. J Eng Geol. 2004;12(1):63–8.

Qi SW, Wu FQ, Liu CL, Ding YH. Engineering geology analysis on stability of slope under earthquake. Chin J Rock Mech Eng. 2004;23(16):2792–7.

Qiao JP, Pu XH, Wang M. A study on characteristics of distribution of earthquake-induced landslides and hazard zoning. 2009;24(21):25–9.

Qu HL, Zhang JJ, Wang FJ. Seismic response of prestressed anchor sheet pile wall from shaking table tests. Chin J Geotech Eng. 2013;35(2):313–20.

Rodriguez CE, Bommer JJ, Chandler RJ. Earthquake-induced landslides: 1980–1997. Soil Dyn Earthq Eng. 1999;18:325–46.

Sakai H, Sawada S, Toki K. Structure considering tensile failure. 12WCEE, 2000, Paper NO. 678.

Scheidegger AE. On the prediction of the reach and velocity of catastrophic landslide. Rock Mech. 1973;5:231–6.

Seed HB, Lee KL, Idriss IM. Analysis of the slides in the San Fernando dams during the earthquake of Feb. 9, 1971. Berkeley: EERC, University of California. 1973.

Shi C, Zhou JW, Ren Q, Zhou XQ. Ray theory solution of the elevation amplification effect on a single-free-face slope. J Hohai Univ. 2008;36(2):238–41.

Siad L. Seismic stability analysis of fractured rock slopes by yield design theory. Soil Dyn Earthq Eng. 2003;23(3):203–12.

Siyahi BG. Pseudo-static stability analysis in normally consolidated soil slopes subjected to earthquake. Teknik Dergi/Tech J Turkish Chamber Civil Eng. 1998;9(DEC):457–61.

Sun CS, Cai HW. Analysis of landslides triggered by Wenchuan earthquake, seismology and geology. 1997;6(1):25–30.

Sun P, Yin YP, Wu SR, Chen LW. Experimental study Of microstructure and mechanical properties of rocks from Donghekou landslide. Chin J Rock Mech Eng. 2010;29(1):2872–78.

Tao LJ, Su SR, Zhang ZY. Dynamic stability analysis of jointed rock slope. J Eng Geol. 2001;32–8.

Wang, FW, Sassa K. A modified geotechnical simulation model for landslide motion. Landslides-Proceedings of 1st European Conference Landslides. Prague, June 2002, in Press.

Wang HP, Li SC, Zhang QY, Li Y, Guo XH. Development of a new geomechanical similar material. Chin J Rock Mech Eng. 2006;25(9):1842–7.

Wang YH, Cheng WR. Dynamic property of a shaking table simulating earthquake. J Vib Shock. 2010;29(2):99–103.

Wang HY, Xie LL. Effects of topography on ground motion in the Xishan park, Zigong city. Chin J Geophys. 2010;53(7):1631–8.

Wang NQ, Zhang ZY, Wang JD. Forecasting method of sliding distance on typical loess landslides. J Northwest Univ (Natural Science Edition). 2003;33(1):111–4.

Wu XY, Law KT, Selvadurai A.P.S. Examination of the pseudo-static limit equilibrium method for dynamic stability analysis of slopes. Canadian: Canadian Geotechnical Conference. 1991.

Wu ZY, Bao JS, Liu HS. A method for evaluating dynamic safety factor rock slope seismic stability analysis. J Disaster Prev Mitig Eng. 2004;24(3):228–41.

Wu SC, Zhang XP, Liu Y. Analysis of failure process of similar soil slope with weak intercalated layer based on particle flow simulation. Rock soil Mech. 2008;29(11):2989–904

Xie H, Wang SG, Kong JM. Distribution and characteristics of mountain hazards induced by the earthquake of May 12 in Wenchuan, China. J Mt Sci. 2005;26(5):396–401.

Xin HB, Wang YQ. Earthquake induced landslide and avalanche. Chin J Geotech Eng. 1999;21 (5):591–4.

Xu Q, Huang RQ. Kinetics characteristics of large landslides triggered by May 12th Wenchuan earthquake. J Eng Geol. 2008;16(6):721–9.

Xu GX, Yao LK, Gao ZN, Li ZH. Large-scale shaking table model test study on dynamic characteristics and dynamic responses of slope. Chin J Rock Mech Eng. 2008;27(3):624–32.

Xu Q, Chen JJ, Feng WK. Study of the seismic response of slopes by physical modeling. J Sichuan Univ. 2009;43(3):262–6.

Xu Q, Pei XJ, Huang RQ. Large-scale landslides induced by the Wenchuan earthquake. Beijing: Science Press; 2009.

Xu Q, Dong XJ. Genetic types of large-scale landslides induced by Wenchuan earthquake. Earth Sci-J China Univ Geosci. 2011;36(6):1134–42.

Yang CW. Study on seismic dynamic characters of rock slopes and system including formation mechanism of lanslides, stability discrimination of slope and assesment of hazard scope of landslide for slope of bedrock and overburden layer. Doctoral dissertation. Southwest Jiaotong University; 2010.

Yang CW, Zhang JJ, Zhou DP. Research of time-frequency analysis method for seismic stability of rock slope subjected to sv wave. Chin J Rock Mech Eng. 2013;32(3):483–91.

Yang CW, Zhang JJ. Landslide responses of high steep hill with two-side slopes under ground shaking, J Southwest Jiaotong Univ. 2013;48(3):415–22.

Yin YP. Features of landslides triggered by the Wenchuan earthquake. J Eng Geol. 2009;17(1): 29–38

Zhang CH, Pekau OA, Jin F. Application of distinct element method in dynamic analysis of high rock slopes and blocky structures. Soil Dyn Earthq Eng. 1997;16(1):385–94.

Zhang P, Chen XM, Wang XD. Analysis of near-fault ground motion and seismic landslide failure mode in Wenchuan earthquake. J Nanjing Univ Technol (Natural Science Edition). 2009;31 (1):49–55.

Zhang JJ, Han PF. Displacement-based aseismic design method for gravity retaining walls-large scale shaking table tests. Chin J Geotech Eng. 2012;34(3):416–23.

Zhang JJ, Liao Y, Qu HL. Seismic damage of earth structures of road engineering in the 2008 Wenchuan earthquake. Environ Earth Sci. 2012;65:987–93.

Zhang JJ, Yang CW, Zhao J, Graeme HM. Empirical models for predicting lateral spreading with considering the effect of region seismicity. Earthq Eng Eng Vib. 2012;11(1):121–31.

Zhao J X, Zhang J, Asano A. Attenuation relations of strong ground motion in Japan using site classification based on predominant period. Bull Seismol Soc Am. 2010;100(5B):2357–80.

Zheng YR, Ye HL, Huang RQ. Analysis and discussion of failure mechanism and fracture surface of slope under earthquake. Chin J Rock Mech Eng. 2009;28 (8):1714–23.

Zhong DH, An N, Li MC. 3d Dynamic simulation and analysis of slope instability of reservoir banks. Chin J Rock Mech Eng. 2007;26(2):360–7.

The manufacturer's authorised representative in the EU is Springer Nature Customer Service Centre GmbH, Europaplatz 3, 69115 Heidelberg, Germany. If you have any concerns regarding our products, please contact ProductSafety@springernature.com

Printed and bound by CPI Group (UK) Ltd, Croydon, CR0 4YY
30/09/2026
02225498-0006